Climate Policy

THE LEADING INTERNATIONAL, PEER-REVIEWED JOURNAL ON RESPONSES TO CLIMATE CHANGE

S0-BRI-485

Climate Policy is a leading journal of analysis and debate on responses to climate change. It addresses both the mitigation of, and adaptation to, climate change. It also provides a forum for the communication of research, analysis, review, and discussion of responses to climate change, including issues related to the UN Framework Convention on Climate Change, the Kyoto Protocol and the negotiation of associated policy instruments. The journal makes complex, policy-related analysis of climate change issues accessible to a wide policy audience and facilitates debate between the diverse constituencies now involved in the development of climate policy.

TOPICS INCLUDE

- Analysis of mitigation or adaptation policies
- Studies of implementation and prospects in different countries and/or sectors
- Applications of integrated assessment to specific policy issues
- Policy and quantitative aspects of land-use and forestry
- Design of the Kyoto mechanisms
- Analysis of corporate strategies on climate change
- Socio-political analysis of prospects for the UNFCCC regime
- Economic and political aspects of developing country action and involvement
- Social studies of climate change, including public perception, where policy implications are derived

Visit www.climatepolicy.com for further information about this journal

JOURNAL SUBSCRIPTION INFORMATION
Institutional subscriptions (6 issues) are £320/$595/€464 per volume (air mail extra); personal subscriptions are £99/$185/€144 per volume (air mail extra)

Abstracting/indexing: ISI social sciences citation index

EARTHSCAN

Earthscan

2 Park Square, Milton Park, Abingdon, Oxon OX14 4RN

Simultaneously published in the USA and Canada by Earthscan

711 Third Avenue, New York, NY 10017

Earthscan is an imprint of the Taylor & Francis Group, an informa business

Climate Policy 5 (2005) 243

Preface

The European Forum on Integrated Environmental Assessment (EFIEA) is a network of research organizations across Europe, brought together through a Concerted Action funded for the period 2002–2005 under the Fifth Framework Programme of the European Union. The overall purpose of EFIEA is to strengthen the EU science–policy interface by applying integrated assessment techniques through bringing together leading researchers, stakeholders and policymakers to jointly address climate change and energy, transport, land use and water policies in a integrated fashion. One specific objective of the network is to create an effective conduit between the scientific, policy and stakeholder communities at the European level and to explore and make explicit the added value of integrated assessment approaches in a variety of environmental policy areas.

As part of this process, EFIEA commissioned the Netherlands Environmental Assessment Agency (NEAA/RIVM, The Netherlands) and the Tyndall Centre for Climate Change Research (UK) to organize and convene two 2-day workshops with the objective of supporting the EU process of developing a post-2012 climate change policy position through deploying insights and approaches from the field of integrated assessment. The programme for the workshops was developed with the assistance of a Scientific Steering Committee. The two workshops brought together over 50 (mostly) European scientists, policymakers, and representatives from NGOs and industry in order to meet this objective.

The first of the two workshops addressed the question: *What long-term policies for climate change adaptation and mitigation should Europe pursue to adequately enhance sustainability on a European (and global) level?* The second workshop built on the outputs from the first workshop and addressed the question: *What are the policy implications of the elements of a European climate strategy that were identified in workshop 1, for the design of a global post-2012 climate regime? What are the implications of views of other important parties for negotiations on a global post-2012 regime?*

Over the course of the two workshops, ten articles by European scientists, well versed in the tools and approaches provided by integrated assessment, were presented to facilitate discussion on these questions and to address the implications of EU policy within a global regime. Following substantial discussion at both workshops, eight of the authors agreed to revise their articles and submit them for peer review so that they could form part of this special supplement of *Climate Policy*. An opening synthesis/conclusions article (Winne et al., 2005, this volume) provides an overview of the issues that emerged from the twin workshops.

We thank EFIEA – and therefore its funder the European Commission – for sponsoring this special supplement and Michael Grubb and James & James/Earthscan for making this publication possible. Needless to say, the ideas presented here have benefited from the varied inputs from all the workshop participants and we thank them for their time and willingness to share and sharpen ideas. The reviewers of the final articles also brought additional wisdom and insight into this process.

As we move into a new era where the Kyoto Protocol is in force and the serious work of designing the next stage of managing climate change on our planet beyond 2012 is undertaken, we offer these articles as a contribution from Europe to the debate.

March 2005

Mike Hulme, Tyndall Centre for Climate Change Research and UEA, UK,
Bert Metz, Netherlands Environmental Assessment Agency/RIVM, the Netherlands.

Climate Policy 5 (2005) 244–250

Towards a long-term European strategy on climate change policy

Sarah Winne[1,2]*, Alex Haxeltine[1,2], Wouter Kersten[3], Marcel Berk[3]

[1] Tyndall Centre for Climate Change Research, Norwich, UK
[2] School of Environmental Sciences, UEA, Norwich, UK
[3] Netherlands Environmental Assessment Agency/RIVM, Bilthoven, The Netherlands

1. Introduction

One of the key challenges facing European policymakers today is developing a politically credible and economically progressive post-2012 climate policy regime. Over the course of two workshops organized by NEAA/RIVM and the Tyndall Centre in autumn 2004, ten articles were presented by leading European scientists to facilitate discussion on Europe's role in this changing international climate policy arena. The timing of the workshops came at the start of the process of intensifying the intergovernmental role of science in shaping and guiding the climate debate after 2012. These workshops were planned as a forum for a wide-ranging discussion on most of the key topics.

The recent Gleneagles G8 Summit and associated communiqué on climate change provides an important anchor for assessing how a wider and more connected science can contribute to this crucial policy debate in the next 2–3 years. In fact the substantive material covered in the two workshops, and presented in the articles here, relates closely to the points raised in the Gleneagles communiqué. Climate change was, as expected, right at the heart of this global summit, additionally bringing in China, India, and Brazil as the strong and potentially high energy-consuming and carbon-emitting economies of the future. The absolute imperative of working towards such a comprehensive global solution was clearly highlighted during last autumn's workshops. The importance of coalition development was an issue raised both at Gleneagles and during last autumn's workshops. The planned G5 and G8 meeting in November is a vital next step in continuing the Gleneagles process. Technology, and the possibility of increasing the speed with which new climate-friendly technologies are developed and transferred to all countries, emerged as a key focus of workshop discussions. This topic is also raised in the communiqué, which indicates that further action needs to be taken to 'promote innovation, energy efficiency, ... , regulatory and financing frameworks; and accelerate deployment of cleaner technologies, particularly lower-emitting technologies' (*The Gleneagles Communiqué*, point 6a). The communiqué clearly places emphasis on the importance of addressing both adaptation and mitigation. This further implies the necessity of broadening the dialogue about the post-2012 climate regime to include adaptation in the context of sustainable development. The discussions held in the workshops surrounding this topic

* Corresponding author. Fax: +44-1603-591375
E-mail address: s.winne@uea.ac.uk

made it clear that whilst many agree on the importance of both adaptation and sustainable development, there is still much uncertainty about how they can be combined successfully in specific policy measures.

The recent climate pact between the USA, Australia, India, China, South Korea and Japan, known as the Asia-Pacific Partnership on Clean Development and Climate, again shows the effectiveness of the USA in building climate coalitions, and highlights the problem faced by the EU in providing an appealing perspective for post-2012 climate policies to the rest of the world. The USA is effective because it focuses on the needs of developing countries to grow and develop technologies. The EU advocates a 'push and pull' strategy in mitigating emissions because simply 'pushing' technologies is not likely to be effective. However, the EU has not done much practically in developing policies in the 'push' component. The new pact makes it clear that the EU needs to acknowledge the dilemma of developing countries to reconcile their economic ambitions with climate protection and to develop clear international policies in the area of supporting climate-friendly technological development and transfer. The G8 process indicates a change in the US position. While it has not yet resulted in concrete results, the process offers an opportunity for the EU to regain its climate leadership role in shaping progress in international climate policy beyond 2012.

It is critical for the EU to keep to, and then go beyond, the objectives of the Kyoto Protocol, reinforcing internal debates on the potential for achieving this goal. The outcomes of Gleneagles clearly make this analysis even more pertinent. Indeed, the spotlight is again on Europe, as the importance of the EU taking a political leadership role has increased. This special supplement of *Climate Policy* brings together a selection of articles that highlight many of the key issues surrounding the post-2012 climate policy discussion.

2. Scope of the workshops

The workshop articles in this supplement cover a range of topics that are central to the development of a post-2012 climate change policy regime, namely: climate change and sustainable development, technological development and policies, short- and long-term climate protection targets, Europe's position in post-2012 negotiations, and climate change adaptation strategies. These articles have been edited since the workshops to reflect workshop discussions, and have also undergone peer review.

Sustainable development in the context of post-2012 negotiations was a key topic of discussion during the two workshops. Petra Tschakert and Lennart Olsson's article on post-2012 EU climate action in the framework of sustainable development policies offers an interesting perspective on sustainable development. This article raises many questions for future EU climate-change-sustainable development policy, the most important being whether climate change policy should 'piggy-back' on sustainable development policies or not. Other questions addressed are: how can the Kyoto Protocol Clean Development Mechanism be developed to better address sustainability and equity? How can unsustainable consumption and adaptation patterns be addressed? The question of how the EU can build a vision of sustainable development is also explored, identifying obstacles standing in the way of this goal.

The topic of technological development also proved central to workshop discussions, as raised in the article by Cédric Philibert. A number of technology policy issues were discussed, including technical change, behaviour and price, competitiveness, scarcity and price, and learning-by-doing.

The importance of exploring the international dimensions of these issues is clear, specifically international technology collaboration, technology diffusion and transfer, and intellectual property rights. The article demonstrates that there is significant opportunity for the development of low-carbon technologies, and that policies and synergies at the EU level have significant potential to impact these technologies. The question is then, what policies are most appropriate?

The issue of long-term climate protection targets and how these can guide future post-2012 climate policy is raised in an article by Jan Corfee-Morlot, Joel Smith, Shardul Agrawala and Travis Franck. They explore how scientific knowledge about climate change impacts can be used and how that in turn can facilitate the forming of new coalitions in international negotiations. Christian Azar's article explores the impact that long-term climate change mitigation policies will have on costs and on European competitiveness. The article identifies the problems that occur when only selected regions are acting on stringent carbon emission reduction targets. The article also demonstrates, however, that increased carbon targets are compatible with increased global and regional economic welfare. Måns Nilsson and Lars Nilsson explore the need for and the difficulties in developing an integrated European policy agenda that works within different sectors and helps to build an integrated international process.

Europe's unique position in climate policy negotiations is discussed in Frank Biermann's article, which questions the role that Europe might play in the emerging coalition formation. Europe has the potential to be drawn into the conflicting interests in climate governance, notably between the USA and the developing world. Lastly, the issue of adaptation is discussed in articles by Farhana Yamin and Frans Berkhout. These articles raise several questions, including: What are the present adaptation challenges to all parties? What are the challenges specifically for the EU? What is the balance between mitigation and adaptation? The articles also indicate areas that need to be addressed immediately, including impacts on the communities themselves, the possibility that 'past' knowledge may not necessarily be relevant to deal with future scenarios, and incorporating the cycle of learning-by-doing into adaptation work.

3. Workshop insights

The conclusions summarized below are the most important broad areas of consensus that arose during *both* workshops. Participants agreed that these points must be addressed when thinking about post-2012 climate change policy from the perspective of the EU, even though agreement on more specific details or implementation strategies was not always reached. The summaries therefore include points of both agreement and disagreement and elaborate on the issues that will need to be addressed in future climate policy discussions.

3.1. Set conditions and processes internally (e.g. EU institutional development) and externally (coalition development)

Thinking about the issue of climate change must be expanded so that it is considered as more than simply an environmental issue. There is a need to engage more sectors both nationally (such as economics, development and foreign ministries) and within the EU (such as a wider range of Directorate Generals than just environment). The EU has the potential to act as a leader in the post-2012 climate change debate, but it must improve coordination and communication internally to do so effectively. Indeed, some of this integration is already beginning to happen. However, interdepartmentalism, especially across national governments, does not come easily when the policy arena is fuzzy and the time-scales are long, and early actions are difficult to justify. This is a research arena of great moment for political scientists.

Europe needs to pay attention to partnerships that are being built, especially with partners in the South and with the USA. This requires: (1) better understanding of the motivations, drivers and standpoints of others; (2) increasing credibility by full implementation of Kyoto; (3) having a position about the continuation of European policies (e.g. what is Europe's position on its own post-2012 targets?); (4) making use of the current dynamic and windows of opportunity to put post-2012 issues on the agenda; and (5) getting high-level policymakers and stakeholders together to develop shared visions and coordinate policies (e.g. governments should engage more in public–private partnerships). Once again, the inherent problems of thinking and action over generations, when the stakes of engagement are so ambiguous and the payoffs so very unreliable, will require the attention of experts in many scientific disciplines working in close harmony.

3.2. A diverse range of targets and commitments will be necessary both within the EU and globally

In general, a more sophisticated way of setting targets is needed compared with the approach in the Kyoto Protocol. There is much debate over whether the post-2012 agreements should include binding targets on reduced emission levels for certain countries. Because such targets are not feasible for all countries, the question needs to be asked – What will other countries be willing to take on over time? Intensity and other indexed targets, and non-binding targets, though still controversial, could provide a way toward allowing developing countries to engage, encouraged by the possibility of participating in emissions trading.

However, it needs to be said that emissions trading depends on ever-tightening 'caps' which convey higher costs and more effort and attention by emitters. The lobbying by business, national governments and aggressive civil economic groups will intensify. This is another arena in which social scientists must work with the modellers.

The importance of addressing the ultimate objective of the UNFCCC ('avoiding dangerous interference with the climate system'), which the EU has translated into its 'not more than 2 degrees' target, has proven to be an important one, because long-term targets provide guidance for short- and medium-term policy. Post-2012 climate policy in the EU and in EU negotiating positions should be consistent with the '2 degrees' target and, therefore, the EU should mobilize as much support for this target as possible. There is a clear need for continued research into climate change impacts, and to engage others in thinking through the implications for emission reduction.

3.3. Enhance technological change

Many issues dealing with technological change and policies have proven complicated; however, one point that emerges is that the diffusion of clean technologies must be accelerated. With regard to technological sharing, one possible idea is to create a mechanism for increasing financial incentives for sharing.

The role of the government in promoting clean technologies is a contentious issue which raises the question as to whether governments should pick 'winning' technologies, or instead choose technology-neutral instruments. Debate continues over whether it is necessary for policies to promote specific technologies, so that they can enter the market and be fully utilized.

The positive aspects of technological change could be emphasized in future debates so that we can take advantage of the current awareness of the benefits of modern technology. This would move the discussion away from the negative aspects of mitigation, towards technology development and business opportunities. In addition, a combination of technology 'push' and 'pull' is needed. Technology 'push' policies will be useful, in particular for the long term, but pulling new technologies into the marketplaces will require price signals stemming from comprehensive instruments such as taxes or cap-and-trade regimes.

3.4. Kyoto Protocol features such as the flexible mechanisms will need to be retained, and expanded upon, in a post-2012 regime

The success of the Clean Development Mechanism (CDM) is an issue that has been debated significantly and has received mixed reviews. Rather than participating in 'CDM-bashing' (because of the current bureaucratic procedures and the small scale), the EU needs to support the CDM market and let the CDM deliver. In addition, emissions trading should certainly be retained in a post-2012 regime, and if possible be expanded.

3.5. Adaptation and sustainable development need to be addressed

The link between climate change and sustainable development needs to be clarified. A post-2012 climate policy regime could be created without addressing sustainable development explicitly, although for developing countries this linkage might be crucial; the reverse (i.e. addressing SD without addressing climate change) is not possible. There is debate as to whether it is necessary to make the link between climate change and sustainable development explicit in order to make progress. Climate change is certainly a sustainable development issue and there are important policy linkages, but diluting climate change in the broader SD context might slow progress on climate change. On the other hand, singling out climate change might not attract sufficient policy attention in many countries. Inevitably, the connection between the two will test the theme of cooperative intergovernmentalism that is proving so difficult to address. Here again, this is an arena where a wider basis for science engagement will prove necessary.

There has been much discussion regarding the need to link adaptation with impacts and vulnerability. In the last decade the focus within climate change negotiations has been too much on the costs of mitigation: there is a great need to complement this with attention to impacts and adaptation, as well as to the link between avoiding impacts and required mitigation strategies within the EU and externally. Certain elements of the post-2012 negotiations, including adaptation, could be based on the FCCC rather than on the Kyoto Protocol, because not all countries have signed the Kyoto Protocol. In the context of adaptation, it may also be constructive to make use of non-FCCC instruments such as existing international disaster relief agreements.

3.6. Additional insights and conclusions

Issues of European costs and competitiveness must be addressed when considering post-2012 policies. The timing and management of transitions to sustainable economic practices (in energy, industry, land use, transport, agriculture and forestry) are important in order to limit costs. Climate change policy options cannot be properly evaluated based on traditional cost–benefit analysis because there are too many uncertainties. Least-cost approaches, i.e. achieving targets at the lowest possible costs, are usually preferable. The use of price caps could provide a way of reducing the fear of excessive costs, while advancing towards significant emission cuts. However, there are many complications in making such a system work.

Both short- and long-term goals need to be addressed when developing post-2012 climate policy. There is need to link long-term goals and visions to short-term action. The choice of long-term targets does matter for short-term policy actions: the next decades will be decisive in keeping options open for staying below the 2 degrees target in the future. To make this work, there needs to be better insight into what changes are needed in specific sectors, and when these changes should occur. Short-term opportunities and critical policy and investment decisions to get on another trajectory need to be identified. The roles of the governments, private sector and the role of NGOs need to be explored further.

There are several areas of research that are of critical importance and need further work. Specifically, there is a need to strengthen strategic research on (a) costs of inaction and the benefits of climate policy; (b) costs and consequences for individual sectors; (c) options for dealing with competitiveness problems; (d) regional climate change impacts in relation to extreme events; (e) coping capacities and (limits to) adaptation and adaptation costs; (f) ways to better integrate climate change into sector and structural policy decisions; and (g) perceptions and social acceptability of climate impacts and policies.

4. Conclusions

These two workshops successfully brought together policymakers, scientists, representatives from NGOs and industry to discuss Europe's role in developing a post-2012 climate change policy regime. Altogether, 58 people participated in the workshops and added their expertise to the discussion.

The discussions that took place during the workshop are especially relevant in the light of the recent Gleneagles Summit, the new USA-led climate technology initiative, and looking to the increasingly intense international debate on the future of the climate regime after 2012. We therefore summarize the key insights into future science-policy discussions as follows:

- There is now a major debate on the future of the Kyoto versus non-Kyoto mechanisms. Scientific research and assessment on climate change must engage with this in a balanced way. Science can contribute to the exploration of what might be the elements of, and pre-conditions for, adequate institutional frameworks for addressing climate change.
- The role of technology in tackling climate change is a key issue. These workshops demonstrated that exciting new research on induced technological change and on the diffusion of new technologies is increasingly able to provide insights into how to realise the potential contribution of technology.
- There are important debates on timing, trajectories and the role of trading. The Gleneagles statement challenges the scientific community to continue to demonstrate the extent to which the science justifies action in the short term and to show how trading (and technology) can reduce the costs of early action.

- A debate is emerging surrounding policies that pay for adaptation. It appears that this may become increasingly attractive, especially where such policies can be seen as stimulating economic growth and contributing to development objectives. However, such policies must be carefully balanced against investment in mitigation in order to avoid perverse incentives. Scientific research is needed in order to first expose this point and then to explore how such adaptation–mitigation linkages could work in practice.
- Finally, the importance of a sustained dialogue between groups of policy-makers and scientists (including both social scientists and climate scientists) emerged clearly from these workshops. This requires personal commitment from both sides to build up a common knowledge base over time. Without this it is very difficult to achieve the depth of discussion required to really add value to policy debates. This demands of the scientists involved that they are cognisant of the subtleties and pace of the policy debate, rather than being wedded to narrow scientific research agendas.

There is certainly scope for future workshops of this kind, although it is essential that they should now be highly focused on particular issues of direct current relevance to the policy debate (such as the issues highlighted above) and, where possible, anticipate emerging 'hot topics' in the policy debate. To this end a consortium of the institutions involved in these workshops is now planning to organize a series of 6-monthly science-policy workshops in Brussels during the crucial 2006–2008 period. This is an important part of the activities of ADAM, a new European research project on long-term adaptation and mitigation strategies for Europe. We specifically envisage the added value of such dialogues as resting on the building of personal contacts and a common knowledge base over time. Such dialogues should play an essential underpinning role in allowing the European Union to provide a critical part of the leadership on climate change that will be essential to making significant progress.

References

The Gleneagles Communiqué, G8 Gleneagles, 2005. Climate Change, Energy and Sustainable Development [available at http://www.number-10.gov.uk/files/pdf/Gleneagles_G8_finalcommnique.pdf].

Asia-Pacific Partnership on Clean Development and Climate, 28 July, 2005. [Available at http://www.dfat.gov.au/environment/climate/050728_factsheet.html].

Climate Policy 5 (2005) 251–272

Long-term goals and post-2012 commitments: where do we go from here with climate policy?

Jan Corfee Morlot[1]*, Joel Smith[2], Shardul Agrawala[3], Travis Franck[4]

[1] Department of Geography, Environment, Science and Society Research Programme, University College London, UK
[2] Stratus Consulting, Boulder, CO, USA
[3] OECD, Environment Directorate, Climate Change Programme, Paris, France
[4] MIT, Joint Program on the Science and Policy of Global Change, Cambridge, MA, USA

Received 17 January 2005; received in revised form 11 March 2005; accepted 18 March 2005

Abstract

With entry into force of the Kyoto Protocol in 2005, climate change negotiators are turning their attention to the question, 'Where do we go from here?'. A key component of answering this question is in understanding the implications for society of alternative long-term goals for greenhouse gas concentrations. One challenge in ongoing negotiations is whether and how to deal with meanings of 'dangerous interference' as outlined in Article 2 of the UN Framework Convention on Climate Change. This study addresses Article 2 by suggesting the use of long-term goals to guide decisions about the stringency and timing of future climate change commitments. Focusing on mitigation policy benefits and, in particular, on avoiding long-term climate impacts, a number of management approaches and their implications are highlighted. After discussing some challenges of using scientific knowledge to monitor and manage progress, we look at what we can learn from current climate change global impact literature. Solid benchmark indicators appear to be available from global mean temperature change, ecosystems and coastal zone impacts information. We conclude by arguing for global goal-setting based on climate change effects and the use of indicators in these areas as part of post-2012 climate change negotiations. Aggregate global impacts suggest that 3–4°C of global mean temperature increase by 2100 (compared to a reference period of 1990) may be a threshold beyond which all known sector impacts are negative and rising with increasing levels of warming. However, marginal benefits may accrue at lower levels of mean change. Thus, a prudent policy might aim for significantly lower levels and slower rates of global warming.

Keywords: Climate change; Science and policy; Global environmental change; Climate change impacts; Greenhouse gas mitigation policy; Climate change mitigation and adaptation linkages

---

* Corresponding author.
E-mail address: j.morlot@ucl.ac.uk

Decision making has to deal with uncertainties including the risk of non-linear and/or irreversible changes and entails balancing the risks of either insufficient or excessive action, and involves careful consideration of the consequences (both environmental and economic), their likelihood, and society's attitude towards risk (IPCC, 2001a).

1. Introduction

Article 2 of the United Nations Framework Convention on Climate Change (UNFCCC 1992)[1] focuses international effort on avoiding dangerous climate change and stabilizing atmospheric concentrations of greenhouse gases. In this light, the consideration of long-term goals to guide policy is not an option but an integral part of the Convention and of future negotiations. Yet a practical interpretation of Article 2 has so far eluded policymakers. Some observers have suggested there is a growing acceptance and political recognition that information about climate change impacts can help policymakers to interpret Article 2 (Oppenheimer and Petsonk, 2005; Yamin and Depledge, 2004; Corfee Morlot and Höhne, 2003; Metz et al., 2002; Berk et al., 2002), while others remain more sceptical (e.g. Pershing and Tudela, 2003). However, much of the research community, and the Intergovernmental Panel on Climate Change in particular, has avoided getting into this area, noting that it is the work of policymakers (rather than scientists and researchers) to balance different perspectives on risk, to value risk avoidance, and to make judgements about what is acceptable (IPCC 2001a; Agrawala, 1999).[2] The formal political process of international climate negotiations under the Convention has also largely avoided the question of how to interpret Article 2 (Corfee Morlot and Höhne, 2003; Depledge, 2000).[3]

Meanwhile several national governments have made hortatory statements about long-term objectives for climate policies; e.g. the Netherlands and, more recently, Germany and Canada (for a detailed, historical account, see Oppenheimer and Petsonk, 2005). In 2005, the EU formally reaffirmed its view on the Convention objective by stating that global mean temperature should not exceed a 2°C increase above pre-industrial levels (EU, 2004, 2005). In an effort to influence an upcoming G-8 ministerial, a multilateral task force of prominent scientists and policymakers recently recommended the establishment of the same long-term temperature change goal as a guide for further policy actions (ICCT, 2005). These actions demonstrate an ongoing political interest in some countries, regions and among some communities of experts to interpret the Convention's objective in a practical way. However, given the global nature of the climate change problem, none of these efforts will have political implications for next steps under the Convention unless relevant issues are also addressed in a wider multilateral context.[4] This article explores the possibility of using long-term goals as a guide for post-2012 climate policy. It focuses on global mitigation decisions and considers long-term climate goals in the form of impacts (or proxies for impacts) that society might wish to avoid. Such goals could be used to outline desirable global emission pathways of acceptable stringency and sufficient timing to significantly limit the risk of key impacts (Figures 1 and 2). For example, distinct pathways emerge for 450 and 550 ppm CO_2 stabilization objectives (Figure 2) and each corresponds to a different set of risks for climate change and impact outcomes. While still not providing insights on important questions related to post-2012 climate negotiations, such as allocation of responsibilities for mitigation among different Parties, forms of commitment, or instruments for implementation, a general agreement on long-term goals would provide input with respect to level of ambition and upper bounds for emission pathways for any future agreement.

This article focuses narrowly on the implications of our current knowledge of climate change impacts for the stringency and timing mitigation commitments. After a brief review of recent IPCC findings of relevance, it is organized into four parts, addressing the following questions:

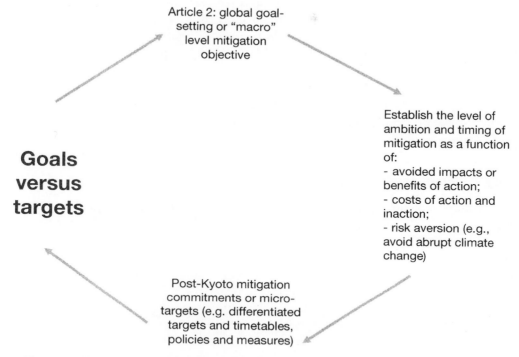

Figure 1. Goal versus target setting for global climate policy.

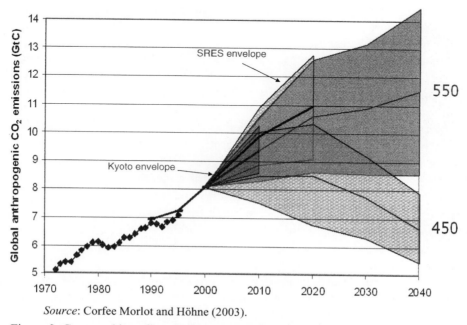

Source: Corfee Morlot and Höhne (2003).

Figure 2. Costs and benefits of climate change policies across time and space.

- What are the main challenges to advancing a policy discourse on Article 2?
- How might we use current knowledge, including the global climate impacts literature, to guide interpretation of Article 2? In particular, what types of climate change impacts (or metrics for impacts) might be used to guide policy decisions on both mitigation and, to some extent, priorities for adaptation?
- Are there transferable frameworks/approaches from other environmental regimes that emphasize decision-making targets based on adverse impacts?
- How might information on impacts be better used to assist in the post-2012 climate change negotiations, advance policy decisions, and review progress towards achievement of Article 2?

2. Advancing policy discourse on Article 2: some analytical challenges

The Intergovernmental Panel on Climate Change (IPCC) has offered some insights into how one might monitor progress towards Article 2. On the question of the benefits of mitigation, the IPCC concluded that 'comprehensive, quantitative estimates of the benefits of stabilization at various levels of atmospheric greenhouse gases do not yet exist' (IPCC, 2001a, p. 102).

A range of different perspectives on the benefits of stabilization were expressed by Smith et al. (2001) in the IPCC summary chapter on impacts. On economic benefits, they summarized literature containing aggregate monetized estimates deriving from integrated assessment exercises, representing gains or losses in global gross domestic product associated with different levels of global mean temperature increase from 1990 (IPCC, 2001c, pp. 943–944, Fig. 19–4). Unfortunately such estimates have a number of important shortcomings: some of these studies ignore significant ecosystem and amenity values or other non-market impacts of climate change; they exclude treatment of the risk of non-linear, abrupt change or shifts in climate variability with implications for extreme climate events.[5] Furthermore, the various authors may use quite different regional aggregation approaches, baselines, and valuation assumptions that drive outcomes, even when controlling for discount rate assumptions. Overall, the results of aggregate studies must be interpreted with much caution because they are few in number, they build on an impacts literature that is not comprehensive in its study of potential impacts, and because they either omit potentially critical impacts or make sweeping assumptions in order to include such impacts. The bottom line is that aggregate damage costs are incomplete at best and need significantly more research before we can have confidence in any of these estimates for policy assessment purposes.[6]

Smith et al. (2001) also provided the policy community with another vision of how one might consider impacts in a comprehensive manner (IPCC, 2001c, pp. 917–959, Fig. 19–7) using a variety of different metrics – economic and non-economic – for assessment of risks across different 'levels of climate change'. They outline five 'reasons for concern': risks to unique and threatened systems; risks from extreme climate events; distribution of impacts; aggregate impacts; and risks from future large-scale discontinuities. These suggest the use of a range of different metrics and benchmark indicators of 'key vulnerabilities' to work within and across sector impact categories, to further develop a means to monitor change. Eventually such indicators could be used to assess the performance of different mitigation strategies.

Mitigation aside, some amount of human-induced climate change is inevitable, particularly through to 2050, but also beyond 2050. This is due to the long time lag between reductions in emissions, changes in atmospheric concentrations, and thus impacts. A sense of what may be already 'locked in' to the bio-geophysical system can be taken from the range of global mean temperature projections found in the literature (IPCC, 2001a, 2001b; Hare and Meinshausen, 2004). Looking across the full range of IPCC scenarios, we see a relatively narrower range of temperature changes predicted in the 2000–2025 and 2025–2050 time frames compared with 2050–2100 (Allen et al., 2001; IPCC, 2001a). This indicates that mitigation action is likely to have relatively limited effects on nearer-term climatic change. Limiting climate change impacts in the short term will require adaptations to address current (socio-economic) vulnerabilities and efforts to avoid mal-adaptations to climate change (OECD, 2003a–d, 2004a–b; Klein, 2001).

Policymakers could make better use of the climate change impacts literature if it were organized around time frames that can be mapped to policy responses. Near-term and unavoidable impacts require immediate adaptation attention, whereas longer-term impacts can be avoided through mitigation and/or altered by adaptation. A better framing of impact literature – both into relevant time frames and into a limited set of benchmark indicators of key vulnerabilities or thresholds – could thus enhance the value of impacts information for use in the mitigation policy decision-making. The remainder of this article explores this possibility, expanding upon the IPCC's use of multiple-metrics and 'reasons for concern' as a way forward to elaborate the benefits of mitigation policy as an input to goal-setting for global mitigation. We build on a recent proposal by Jacoby (2004; see Box 1) as well as other recent efforts in this area (OECD, 2004c; Corfee Morlot and Agrawala, 2004; Patwardhan et al., 2003; ECF and PIK, 2004; UK Defra, 2005), to begin to carefully consider the links between avoided climate change impacts and possible post-2012 mitigation efforts.

3. Use of Article 2 to guide decisions and monitor progress

Broad uncertainty may explain the lack of attention in the negotiations to the interpretation of Article 2 and long-term goals of climate policies. Uncertainty complicates such an interpretation, especially since the risks of accelerated climate change will play out over decades or even centuries. There is also a disconnect between the location of mitigation action and the location of where policy benefits occur. As impacts are a function of changes in physical climate and socio-economic conditions (including vulnerability and the ability to deal with climate change), the principal benefits of mitigation policies are not likely to accrue to those who mitigate most. Despite these problems, some guidance on how to approach monitoring or 'operationalization' of the Convention objective is found in Article 2 itself, which outlines three key vulnerabilities to guide decisions: stabilization should be achieved in a time frame (i) to avoid threatening food production, (ii) to allow ecosystems to adapt naturally, and (iii) to enable economic development to proceed in a sustainable manner.

The three vulnerabilities mentioned in Article 2 may be considered benchmark impact areas against which progress through international cooperation can be monitored. Although monitoring progress in this way has not yet occurred, it may be an appropriate time to begin. Such a 'stock taking' may be a valuable starting point for new negotiations on post-2012 commitments, especially if those negotiations are to include a broad understanding of implications of their choices for long-term outcomes or impacts. The notion of key vulnerabilities or thresholds in different time frames may provide insights. Consistent with the Jacoby proposal (Box 1), two basic approaches to understanding key vulnerabilities are discussed here: 'top-down' approaches using global proxies for climate change, and 'bottom-up' approaches using regional impact assessments and information.

Box 1: Avoided climate impacts: a portfolio of measures to consider links with mitigation decisions

Jacoby (2004) proposes **a portfolio of different measures** to provide a structure for current (incommensurable) information on avoided impact benefits of mitigation and future research. It starts with information on **global physical variables** which are characterized with quantified uncertainty ranges so as to cast the issue in risk reduction terms. A selection of global variables (e.g. mean temperature, perhaps by latitude or sea level) should be expressed in natural or physical units. This level of assessment would be free of valuation or aggregation controversies, as the information is reported separately by variable and in different physical units. A priority for research at this level would be on how to characterize and communicate with stakeholders on how to deal with different types of risk associated with different levels of climate change. Also at this global scale, at least qualitative characterization of abrupt, non-linear change should be included (e.g. Schneider and Lane 2004; Alley et al., 2003).

A second set of information is the **characterization of effects at regional scale.** This might be mainly natural/physical units for non-market impacts and natural/physical units plus monetary estimates for market impacts where possible. A first step is to rethink the regional groupings to be used to structure the information. Ideally the structure would include only a few indices which would have high information content to a wide body of stakeholders. These should be clearly defined and allow for independent assessment with global applicability (across regions), ultimately allowing for comparison across regions.

A third set of information is bottom-up, regional-scale **non-market impact valuation and global aggregation** which would allow valuation to proceed as methods and data allow. This leaves aggregation in the hands of the individual players, and provides clarity about different approaches to valuation, weighting and aggregation.

Source: Jacoby (2004).

3.1 *'Top-down' thresholds to structure assessment and guide decisions*

One possible proxy for global climate change impacts in mitigation analysis is global mean temperature change (Jacoby, 2004). Characterization of impacts as a function of global mean temperature increase[7] could facilitate review of mitigation options and of trade-offs among different strategies with respect to climate change impacts and costs (Jacoby, 2004; Hitz and Smith, 2004; Corfee Morlot and Agrawala, 2004). This is because global mean temperature increase will vary directly with changes in atmospheric concentrations of GHG. In this way, global mean temperature increase (over a common base year period) can be a means to compare the implications of different 'levels of climate change' over time.[8]

Recent literature has begun to quantify uncertainties in GHG emission projections and climate predictions (Webster et al., 2003; Jones, 2004; Wigley, 2004a and 2004b; Wigley and Raper, 2001). These probabilistic assessments could be a useful guide to establishing upper bounds for global emissions or 'emission envelopes' for future mitigation decisions.[9]

Through the use of probability distribution functions, one can look at the likelihood of exceeding a particular global mean temperature change under different atmospheric stabilization outcomes or across emission scenarios. In this approach the question for the policymaker becomes: *What is the probability of exceeding a particular change in global mean temperature over 1990 levels (say 2–3°C) for one particular mitigation strategy versus another* (Jones, 2004; see Figure 3)? An

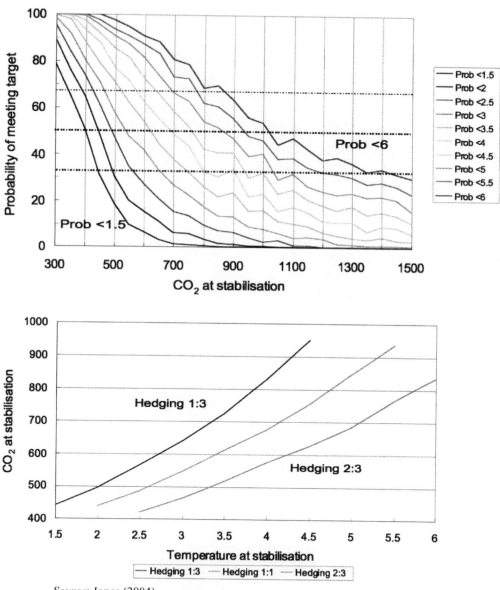

Source: Jones (2004).

Figure 3. Probabilities of meeting temperature targets at given levels of CO_2 stabilization. For a given level of atmospheric CO_2 stabilization in the future, a range of climate variables (e.g. forcing constants and climate sensitivity) were varied. Jones (2004) performed uniform Monte Carlo sampling of these variables and then, using an equation detailed in his paper, calculated the stabilization temperature.

alternative way to phrase the question is: *What is the distribution of global mean temperature increase outcomes (and its most likely value and confidence interval), associated with any particular mitigation strategy* (Webster et al., 2003; Jacoby, 2004; see Figure 4)?[10] Approaches to answer either of these questions are likely to combine expert opinion and analytical modelling to derive likelihoods of the outcome in question through probabilistic assessment (e.g. Reilly et al., 2001; Allen et al., 2001; Pittock et al., 2001).

The top-down 'threshold' approach embedded in the first question presumes the existence of (and agreement of the policy community on) some upper bound of global mean temperature change which is to be prevented from occurring. A shortcoming of basing the assessment on such a physical threshold is that it could lead to the interpretation that prevention of lesser levels of change is not of value. However, an advantage is that it reveals clear differences in the substantive outcomes of one mitigation strategy versus another when taking into account recognized uncertainties.[11]

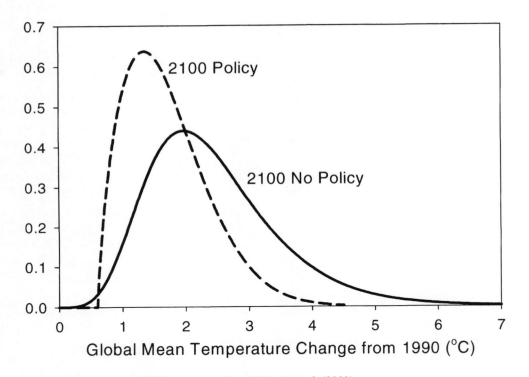

Source: Jacoby (2004) as adapted from Webster et al. (2003).

Figure 4. Temperature change under no-policy and 550 ppm GHG stabilization cases. Webster et al. (2003) produced this joint probability density function (PDF) by varying five areas of model parameters including climate sensitivity and future GHG emissions. Their model design is such that they can explore structural uncertainties of other models by varying these model parameters. The input PDFs were constructed using historical observations and expert elicitation, and were not uniform as Jones (2004) assumed. The policy case used by Webster et al. (2003) is equivalent to a 550 ppm stabilization scenario using the model's reference parameters.

3.2 'Bottom-up' thresholds to structure assessment and guide decisions

Some authors have also advocated the use of 'bottom-up' impact thresholds to inform decisions about 'what is dangerous' under Article 2 of the Convention (Swart and Vellinga, 1994; Parry et al., 1996; O'Neill and Oppenheimer, 2002). Although this concept has undoubtedly influenced the policy debate in the past, it has not had any tangible negotiating outcome (Corfee Morlot and Höhne, 2003; Oppenheimer and Petsonk, 2005).[12] One reason may be the lack of international consensus on a meaningful set of benchmark bottom-up indicators of global impacts on which one could establish thresholds and monitor progress. Clearly monitoring all impacts would be impractical, if not impossible; instead progress requires agreement on key impacts that are significant and meaningful to both specialist and non-specialist audiences in order to secure broad public support for action.

A second challenge with a bottom-up threshold concept is that many climate change risks and related thresholds are highly context-specific. This requires local or regional assessments of relevant factors, including vulnerability, coping ranges, historical variability and community risk preferences (Dessai et al., 2004; Jones, 2001). Credible threshold assessment involves a complicated set of 'bottom-up' tasks that would be difficult for international negotiators to grapple with directly. However such a 'bottom-up' effort could be commissioned and run in parallel with a formal negotiation to provide a robust base of information as well as greater awareness and support for action to limit climate change.

Despite the challenges, complementing the necessarily top-down negotiating approaches with bottom-up assessments and perspectives may provide a means to bring together over time the widely divergent interests of different negotiating groups around common avoided impact goals. Of course, agreement on a threshold at any scale (global, regional or local) will always require political judgement about what is acceptable and, presumably, this judgement will be grounded in evolving social understandings of climate change. In the case of climate change, the best that one could hope for is a clear indication from certain communities about what thresholds or key vulnerabilities are relevant to whom, which in turn could guide assessment of the performance of alternatives with respect to the identified key vulnerabilities. An example might be ensuring 80% stability in unique and climate-sensitive ecosystems such as Arctic ecosystems (e.g. tundra). An assessment of this key vulnerability could aim to identify benchmark metrics and thresholds, say for regional mean temperature increase or another proxy of global climate change, below which risk to ecosystem resilience is minimized.

Any use of 'bottom up' indicators of regional impacts of climate change, threshold identification or other forms of information gathering on local/regional risk perception would ideally involve an iterative dialogue and scientific assessment process, with engagement of stakeholders in the formulation of recommendations.[13] Some relevant experience may exist in climate change impact programmes already established in OECD countries or regions (e.g. in the UK and in the EU); however, in general, there is limited capacity and experience with the use of regional impact assessments to consider the implications of climate change mitigation strategies over the long term.[14] Including developing countries or other key emitters (e.g. Russia) with OECD nations in such efforts is a long-term agenda, yet one that may pay off by incrementally building support for strengthened mitigation action to limit climate change and its impacts.

At the moment negotiators may have a only a vague understanding of the risks of climate change, which form the basis for negotiating future mitigation commitments, especially with respect to arguments about stringency of commitments.[15] There may already be an important and rapidly growing base of knowledge about observed and predicted climate change impacts (e.g. in the Arctic region, in the Mediterranean, in high altitude mountain regions, and possibly coral reef areas and/or coastal zones). At a minimum, further work and dialogue with stakeholders in regions relevant to these impact areas could build on existing assessments to provide important insights for post-2012 negotiations and possibly provide a glimpse of the implications of alternative mitigation decisions on long-term impact outcomes.

4. Using current global impact literature to interpret Article 2

A recent OECD-led review of the global impacts literature from a mitigation perspective offers some insights about the differences between a world with global mean temperature increases of 2°C, 3°C or 4°C compared with 1990 levels; however, it also raises a number of questions and highlights gaps in our knowledge.

Hitz and Smith (2004) reviewed the global impacts literature to assess the general shape of sector damage curves. They express globally aggregated impacts in particular sectors as a function of changes in global mean temperature (GMT) and in units used in the original studies (i.e. monetary and non-monetary). The analysis considers the magnitude of avoided damage benefits in going from one level of climate change to another. When considering the magnitude of damages, the question arises: *Are changes in impacts constant, decreasing, increasing, or do they change in sign (from negative to positive) at some point?*

The authors conclude that sector damages can be plotted as function of GMT increase – although levels of confidence vary. Damages relationships vary across sectors, however, and these sector variations may provide insights into key vulnerabilities and thus help to guide effective policy responses. Finally, in a number of areas there are no global studies, so our understanding of global impacts remains partial at best.

Hitz and Smith (2004) divided sector damage curves into three categories (Figure 5):

- 'parabolic' in shape (decreasing initially, shifting to increase with more significant climate change)
- increasing with climate change
- indeterminate.

Results showed a parabolic relationship for agriculture, terrestrial ecosystem productivity and forestry – reflecting a positive CO_2 fertilization effect on plant growth at lower levels of warming. At higher temperatures, however, aggregate damages emerge as optimal growing temperatures are exceeded and warmer (sometimes drier) conditions lead to evaporative loss of humidity in soils, thus limiting plant growth.[16] In a number of other sectors – coastal zones, health, marine ecosystems, biodiversity – impacts are negative even at lower levels of global warming. Thus these sectors are likely to be among the most sensitive to climate change at the lowest levels of global warming. Finally, in a number of important sectors – such as water, energy, and aggregated costs across all impacts – the relationship is simply unknown (due to an insufficient number of studies).

No sector results suggest positive impacts from climate change as mean temperatures increase beyond several degrees. Thus global aggregate marginal benefits of mitigation clearly exist beyond global mean temperature increases of 3–4°C (compared with 1990 levels). Though marginal benefits

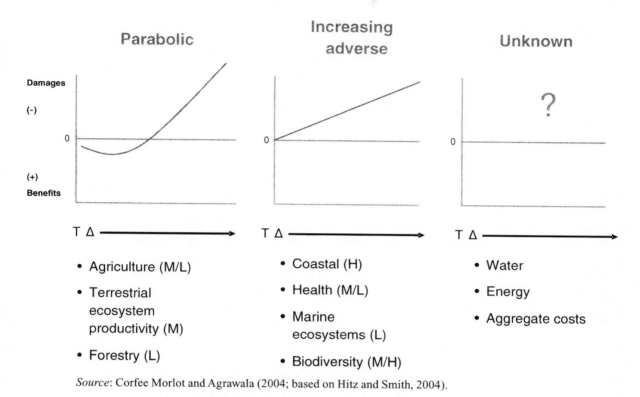

Source: Corfee Morlot and Agrawala (2004; based on Hitz and Smith, 2004).

Figure 5. Sector damage relationships with increasing global mean temperature. These graphs are illustrative only, and do not attempt to fully summarize the variety of relationships that were found in this study. The results are based on global impact assessments only and do not take into account a much larger and richer literature that exists at the regional scale. Also, for some systems/sectors, few studies were available, hence the ranking of uncertainty (High, Medium, Low) associated with each sectoral assessment. No global studies exist for the following sectors/systems: recreation and tourism, transport, buildings, insurance, human amenities.

are likely to accrue at even lower levels than 3–4°C, the literature is somewhat ambiguous, especially in key market sectors such as agriculture and forestry. For ecosystems and for coastal zones, however, marginal benefits will accrue when moving below 3–4°C to no change in climate (from 1990 levels).

What are the policy relevant implications from this survey of global impacts literature? How can such literature help us define 'dangerous interference' as outlined by Article 2? In the agriculture area, evidence (when taken globally) does not point to a stringent long-term target, as it could be argued that small amounts of climate change may be favourable to global agricultural production. Of course, as indicated with the 'medium/low' (M/L) confidence ranking, there is still uncertainty about these conclusions. To raise confidence, agricultural impacts need to be carefully investigated to consider the effect of changes in water supply, potential increases in climate variability, changes in location of agricultural pests and diseases, and other factors. Such a large, regional research agenda is important to informing policy in the coming decades but it is unlikely to produce results in the near term. In addition, we know that food security – the key vulnerability identified in Article 2 of the Convention – is highly dependent on non-climate factors such as socio-economic vulnerability, access to trade and food storage facilities and networks (Cannon, 2002; Fischer et al., 2002; Pingali, 2004).

For these reasons, agriculture and food security more generally may not be the ideal 'bell-weather' indicators to guide policy decisions on climate change in a post-2012 period. Nevertheless it is interesting that even in this sector, where results at lower levels of global warming are mixed if not positive globally, there is an indication of overwhelming vulnerability to negative impacts at 3–4°C of mean average warming. In comparison to agriculture, however, ecosystems and coastal zones appear to be much more sensitive to global warming at lower levels of mean change. Sector results in both of these areas appear to be robust across existing studies and to present risks even in the near term at low levels of global mean temperature change. In this case one might ask: *Which levels and types of risk in these areas are acceptable and which are not?* Coastal zone impacts are important because they directly affect large segments of the world population living near or in coastal areas; as such they could serve as a proxy for risk to sustainable economic development in these regions. Ecosystems are already explicitly mentioned in Article 2 and are explored briefly below to provide an example of the type of guidance that might be derived from the impacts literature for global mitigation policy. Consistent with the IPCC's five 'reasons for concern', we also briefly review recent integrated assessment literature on abrupt climate change in the search to provide guidance for the interpretation of Article 2.

4.1 Biodiversity and ecosystem effects

Studies of aggregate global biodiversity and ecosystem effects are limited and the means for measuring changes driven by climate change are not yet well established. A recent study by Leemans and Eickhout (2004) provides an aggregating methodology by using a series of different indicators of change in ecosystem categories across the world's terrestrial surface area. Global impacts on ecosystems are assessed assuming both low and moderate-to-high levels of climate change (as well as at implied rates of change). Looking across the range of indicators at different levels and rates of climate change, Leemans and Eickhout's conclusions are consistent with the findings of other analyses (Root et al., 2003; Parmesan and Yohe, 2003; Hare, 2005; Etterson and Shaw, 2001) and another model-based assessment of ecosystem impacts (Thomas et al., 2004): even small levels of climate change are expected to have (or already have had) significant impacts on temperature-limited ecosystems, such as the tundra, and on diversity of species within ecosystems.

Leemans and Eickhout also conclude that risks to many regional and global ecosystems rise rapidly above a 1–2°C increase in global mean temperature (by 2100 compared with 1990 levels), mainly due to the inability of forest ecosystems to adapt to such rapid rates of temperature increase. The authors note that mitigation may be the most effective policy to limit ecosystem stresses from climate change, where every degree of avoided global mean temperature increase will yield clear benefits by limiting ecosystem disruption. Relating their assessment back to the five 'reasons of concern' identified by the IPCC, the authors propose adding another reason of concern related to adaptive capacity of regional and global ecosystem (Figure 6).

More detailed system-specific studies (e.g. coral reefs, arctic tundra) could provide better insights into whether system thresholds exist beyond which irreversible change will occur and of how

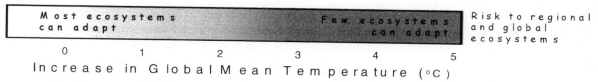

Source: Leemans and Eickhout (2004).

Figure 6. Risk to regional and global ecosystems by global mean temperature increase. Increase in global mean temperature is relative to 1990 estimates (rather than pre-industrial). The main difference between this risk area and the first bar in IPCC's 'burning embers' graph is the focus on ecosystems, whereas 'unique and threatened systems' in the IPCC graph refers both to human systems (e.g. island communities) and to specific and unique ecosystems such as coral reefs or mangroves. More work is still needed to provide a comprehensive understanding of impacts in this area, including decisions on standard metrics for monitoring change, assessing non-linear change in ecosystems and path dependency of such change (see also Schneider, 2004) as well as the economic implications of these changes (Gitay et al., 2001; Smith et al., 2001).

such thresholds are linked to alternative emission pathways and mitigation strategies. It may be that climate change risks to some unique ecosystems, such as coral reefs or mountain glaciers in tropical regions, are unavoidable given the current commitment to future climate change. Here the best strategy to deal with climate change will be adaptive management strategies that aim to boost local resilience and plan for the expected changes in natural systems. Meanwhile the global and regional impact literature cited here implies that preventive or precautionary approaches – aiming to avoid significant disruption to ecosystems – would need to aim at relatively low levels of global warming (i.e. limiting mean global warming increases in this century to 1–2°C above 1990 levels, as well as limiting decadal increase to 0.1–0.2°C). Some authors have suggested that even this may lead to 'dangerous' climate change.[17]

4.2 Risk of abrupt change

A series of high-profile reports have recently brought attention to the risks of abrupt climate change (Alley et al., 2003; Schwartz and Randall, 2003) through low-probability, high-consequence events which have been typically left out of standard impact assessments (Schneider and Lane, 2004; Schellnhuber et al., 2004). Abrupt climate change describes a variety of 'switch and choke' elements in the Earth's bio-geophysical systems that might be activated or deactivated by human interference with the global climate (Schellnhuber et al., 2004). These include collapse of the Amazonian forest, instability of the West Antarctic Ice Sheet, monsoon suppression in the Tibetan Plateau, methane outbursts from permafrost melting, and breakdown of the thermohaline circulation. Analysis of the drivers and thresholds for these 'switch and choke' elements is still in its infancy; however, results from a growing number of studies suggest that accounting for irreversible, abrupt change is likely to shift the economically 'optimal' level of mitigation, calling for more investment in abatement today.[18]

Schneider and Lane (2004) argue that the harder and faster a system is disturbed, the higher is the likelihood of such abrupt events, which could be catastrophic. Similar arguments have been advanced by other experts concerned about the risk of abrupt change (e.g. O'Neill and Oppenheimer, 2002; Oppenheimer and Alley, 2004, 2005). Among the benefits of early and stringent GHG mitigation could be a reduction in the likelihood of such high-consequence events. This conclusion should be explored analytically; if found to be robust, it implies that rate of change of emissions or concentrations may be a useful proxy of climate change to support any long-term goal setting (O'Neill and Oppenheimer, 2004). Most importantly, there has been little study of the consequences of abrupt events. Some, such as the breakdown of the West Antarctic Ice Sheet, would obviously be catastrophic. Others, such as the impacts of changes in the thermohaline circulation, are not so clear.

4.3 Interface between adaptation and mitigation

That some amount of climate change will be unavoidable points to the need to advance adaptation in parallel with accelerated mitigation. Adaptation will require taking projections of medium-term climate change into account in some regions in the planning in infrastructure projects and, more generally, in natural resource management decisions (OECD, 2003a–d, 2004a–b).

A recent OECD case study on Nepal demonstrates the close connections between water resource management and energy production and climate change in a development planning context (Agrawala et al., 2003). Nepal is currently experiencing rapid retreat of the Himalayan glaciers, which has been linked to a trend in rising temperatures (Richardson and Reynolds, 2000; Shrestha and Shrestha, 2004). There has also been an increase in the volumes of glacial lakes, increasing this risk of glacial lake outburst floods (GLOFs). GLOFs may release huge volumes of water that can wipe out any infrastructure and villages that may be in their path. One event of this type in 1985 wiped out a multimillion-dollar hydropower dam that had only recently been constructed; other events have damaged or destroyed roads and villages, adversely affecting the livelihoods of local people. The risk of such events and of glacier retreat more generally alters the water and the energy (hydropower) outlook for the region affected as a whole, and has implications for infrastructure planning and projects (Agrawala et al., 2003; OECD, 2003c).

This example demonstrates the need for a combination of local/regional adaptation and international mitigation strategies to effectively manage regional climate change in high-risk areas. It also demonstrates how impact assessment can point to the limits of mitigation in some areas and to the need for urgent attention to adaptation to address impacts that are unavoidable in future decades, given the known atmospheric changes already locked into the Earth's system.

5. Impacts-based policymaking: lessons from Long-Range Transboundary Air Pollution

Relevant lessons about the use of long-term goals to guide multilateral environmental agreements may exist under other agreements, and the Convention on Long-Range Transboundary Air Pollution (LRTAP) is considered here as one example where scientific information on impacts is used to guide policy decisions (Semb, 2002; Levy, 1993). LRTAP surveys a wide range of pollutants and is organized to generate and share new knowledge on the transboundary environmental effects of these pollutants as well as to manage and limit these effects. The LRTAP process has successfully raised concern and awareness among participating and observer nations about the hazardous environmental effects of acidic pollutants.[19] This has occurred through a 'Working Party on Effects' and the establishment of several (voluntary) international collaborative programmes, which share research on a variety of different receptors of the pollutants. For example, separate programmes exist to study effects on forests, materials, freshwaters and crops. Working in combination with an extensive emissions monitoring programme (EMEP), LRTAP's collaborative approach to developing and sharing knowledge has helped to shape international consensus on the need for political action and to shift a number of initially sceptical or laggard nations to adopt policies designed to curb acid rain (Levy, 2001).

LRTAP has also developed the notion of 'critical loads' to guide decisions on specific multilateral emission limits (Semb, 2002; Levy, 1993). Critical loads define the highest amount of deposition of a pollutant that species or soils can tolerate without changing the ecosystem in an 'unacceptable' way. Though initially resisted, the notion of critical loads for acidic deposition is used by the signatories to this Convention to monitor progress and as an input for decisions on future emission mitigation commitments.

Given both the complexity of the science and its importance to social understandings of climate change and the policy process, future UNFCCC negotiations might be strengthened through an improved science–policy dialogue on post-2012 mitigation options. For example, a useful initiative would be to establish a setting where scientific researchers could usefully interact with, brief and advise international policymakers to inform the debate about the climate change effects of alternative mitigation strategies in the long term. Social scientists might also comprise part of this effort, advising on knowledge about social understandings of climate change. Establishing such an institutional mechanism on a permanent basis would take time but could be instrumental to raising awareness among the policy community, and the participating civil society and business groups, about available expert knowledge. Of course the expert community cannot be expected to pass judgement on what types of risk are politically and socially acceptable. That is the job of politicians – in this case the international negotiators. Nevertheless, by outlining information about the nature of climate change risks, the scientific and expert community can frame the problem in such a way as to help policymakers to interpret Article 2 in an iterative and flexible manner, through incremental decisions across time, building on new knowledge about scientific and social perspectives on climate change risk.

6. Building new coalitions to limit climate change

Inevitably the challenge for negotiators is to build new coalitions that support mitigating climate change and its impacts. Negotiations over the last half a decade have deadlocked due to conflicting interests among diverse participants. In a world characterized by increasingly open trade and investment patterns, and a rise in the influence of the private sector (compared with government), successfully addressing climate change requires governments to establish clear and timely incentives for industry to innovate in order to protect the environment. Delay in providing these incentives can only delay necessary innovation. The limited political will in the international community to broadly regulate GHG emissions among all major emitters indicates a fundamental failure to understand the risk of inaction – that is, the risks of allowing climate change to proceed unabated.

New coalitions could emerge among those nations that have a common interest to limit climate change impacts within this century to avoid risk to natural and socio-economic systems that are vital to local and regional livelihoods (see Figure 7). AOSIS is already such a coalition, but sentiment in these more vulnerable island nations may be joined increasingly by other nations, including those from the 'North', that are also threatened by sea-level rise or other changes in regional climates. This perspective contrasts with that sometimes advocated by others as a way forward – e.g. focusing

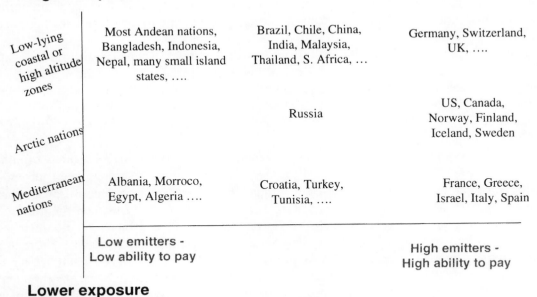

Higher exposure

	Most Andean nations, Bangladesh, Indonesia, Nepal, many small island states, ….	Brazil, Chile, China, India, Malaysia, Thailand, S. Africa, …	Germany, Switzerland, UK, ….
Low-lying coastal or high altitude zones			
Arctic nations		Russia	US, Canada, Norway, Finland, Iceland, Sweden
Mediterranean nations	Albania, Morroco, Egypt, Algeria ….	Croatia, Turkey, Tunisia, ….	France, Greece, Israel, Italy, Spain

Low emitters - Low ability to pay   High emitters - High ability to pay

Lower exposure

Figure 7. Coalitions of nations to avoid the significant risks of climate change.

on the 'large-emitting' countries and allowing them to negotiate the next set of commitments among themselves. If one adds the coalition of highly vulnerable nations to the core negotiating group, the issue of stringency, pathways and upper limits for near-term commitments, such that long-term goals may be achieved, will inevitably be part of the discussion. On the other hand, if the negotiations were to proceed among the largest emitters only, the impact or climate system effects of mitigation decisions could be largely ignored, with the focus remaining on the cost of mitigation alone.

We argue that both perspectives are needed, calling for a combination of highly vulnerable and high-emitting nations to be fully engaged in post-2012 negotiations. The highly vulnerable nations are needed, at least at the outset of such a negotiation, to influence the ambition level or stringency of any near- to medium-term obligations. If there is convergence among the highly vulnerable and high-emitting nations on the global level of ambition of a post-2012 set of commitments, then large emitters might be designated to take over the remaining issues in the negotiations, e.g. to work out the details for the remaining architecture of the agreement.

Underlying this agenda is the need to build the capacity of negotiators to understand the connections between emissions, mitigation, and impacts outcomes. Such capacity could be boosted by a wider understanding and greater public awareness that could be developed through 'bottom-up' regional assessments of critical impact areas (e.g. of ecosystems vulnerable to climate change) and stakeholder dialogues about thresholds or upper bounds for 'acceptable' change. Coral reefs and Arctic ecosystems, and follow-on impacts on human systems, have been cited here as examples of climate-sensitive systems that might benefit from such a bottom-up assessment to better understand the potential for avoided impact benefits of mitigation. Regional assessment combined with stakeholder dialogues might also prove valuable to spread awareness and understanding of both mitigation and adaptation options in heavily populated coastal zones (e.g. S-E Asia, the Mediterranean basin, the Florida Keys) and regions dependent upon water resources from snowfall and glaciers in high altitude mountain systems (e.g. the Alps, the Rocky Mountains, the Himalayas).

7. Implications for post-2012 negotiations

What does an average 2°C warmer world look like compared to a 3°C world, or a 4°C world for that matter? These would seem to be central questions if we are to press for any particular long-term atmospheric concentration target, many of which could suggest stringent and quite radical emission reductions and transformation of our energy economies in the coming decades.

We argue for the use of global goal-setting – based on careful assessment of the long-term impacts of different mitigation goals – as part of post-2012 climate change negotiations. Even a conservative interpretation of the climate change impacts literature, taking into account aggregate global impacts alone, suggests that 3–4°C of global mean temperature increase above 1990 levels by 2100 may be a threshold beyond which all known sector impacts across all regions are negative and rising with increasing levels of warming. Below this, at a global mean temperature increase of 1–2°C by 2100, significant and irreversible changes in ecosystems are predicted, especially if warming occurs at a rapid rate, limiting the ability of ecosystems to adapt naturally. Thus a precautionary policy might aim for significantly lower levels and slower rates of global warming.

The exact level of any long-term goals may be less important to negotiations than discussion of such goals and their implications for regional impacts. Any convergence towards agreement on an upper limit in the long term would indicate upper bounds for nearer-term emissions. The aim of related policy decisions would be to constrain emission pathways such that they leave open the possibility to achieve loosely agreed long-term goals or 'soft targets'. Goal-setting at the outset of negotiations on next steps under the Convention can help to develop a shared understanding of desired outcomes for policies, and upper bounds for global emissions in the nearer term.

Lessons and linkages with international air pollution policy initiatives may point a way forward for climate change. The LRTAP provides an example of international environmental policy driven by the notion of avoided impacts through the use of the *critical loads* concept. *Effects-based* international collaborative research and science policy dialogue, as used under LRTAP, offers one model that could be explored to support reflection on long-term mitigation strategies under the UNFCCC.

Taking Article 2 as central to decisions on next steps could raise awareness within the political process about effects of climate change and its linkages to mitigation in the coming decades. Focusing on climate change effects might also lead to new coalitions of partners (Figure 7). Without a broad focus on environmental effects, *ad hoc* incrementalism may reign, as in the past, to significantly limit the level of ambition and scope of participation in post-2012 mitigation commitments. It is unlikely that basic research in the impacts field will yield new specific results in time to shape decisions on next steps. Nevertheless, raising awareness about what we already know may help to mobilize support for action in developing and developed countries, among policymakers and the general public. It follows that it would also be timely for negotiators to consider the implications of different mitigation strategies for a variety of impact outcomes, in key areas and in different time frames, and to use this understanding to inform their decisions on the architecture for future commitments.

Acknowledgements

The authors have drawn on their work with the OECD on these issues and thank the governments that have supported that work financially, in particular Canada, Finland, Germany and the USA; however, the views in this paper are those of the authors alone and do not represent the views of the OECD nor of the OECD Member countries. The authors would also like to acknowledge comments on an earlier draft from Tom Jones of the OECD Environment Directorate, as well as helpful comments from two anonymous referees on the final draft.

Notes

1   The Article 2 objective of the Convention is: 'to achieve … stabilization of greenhouse gas concentrations in the atmosphere at a level that would prevent dangerous anthropogenic interference with the climate system. Such a level should be achieved within a time-frame sufficient to allow ecosystems to adapt naturally to climate change; to ensure that food production is not threatened and to enable economic development to proceed in a sustainable manner.'

2   The IPCC was asked to address Article 2 in the Synthesis of its Third Assessment Report; however, their response notes that 'scientific evidence helps to reduce uncertainty and increase knowledge…' but that 'decisions are value-judgements determined through socio-political processes…' (IPCC, 2001a, p. 38). The IPCC has not been entirely silent on this issue – holding two workshops on Article 2 (1994 and 2004) – but has made clear that their role is to be limited to providing technical input to policymakers' decisions. See Oppenheimer and Petsonk (2005) for a review.

3   One exception in the international negotiations is in the consideration of the issue of compensation, where the Convention foresees actions [by Parties] to meet the special needs and concerns of developing country Parties arising from adverse effects of climate change (Article 4.8). To help inform these negotiations, the UNFCCC Secretariat has held several workshops featuring recent information on impact assessments (see http://www.unfccc.int – workshop information). These discussions were not, however, related to questions about mitigation commitments.

4   One exception in the international negotiations on compensation that is linked to Article 4.8 of the Convention. This Article foresees actions [by Parties] to meet the special needs and concerns of developing country Parties arising from adverse effects of climate change. To help inform these negotiations, the UNFCCC Secretariat has held several workshops featuring recent information on impact assessments (see http://www.unfccc.int – workshop information). Meanwhile several national governments have made hortatory statements about long-term objectives for climate policies, for example the Netherlands, and more recently Germany, and the EU has formally adopted a view that global mean temperature increase (above pre-industrial levels) should not exceed 2°C.

5   The Nordhaus and Boyer (2000) work includes limited treatment of amenity values and abrupt climate change. Several other exercises have shown that vastly different results can be had by manipulating the way in which abrupt change is characterized (Azar and Lindgren, 2003; Mastrandrea and Schneider, 2004).

6   Nordhaus and Boyer (2000, p.98) also draw a similar conclusion.

7   Note, global mean temperature (GMT) increase can refer to a number of different base year(s) or periods. At least three different reference points can be found in the literature: GMT increase (i) compared to pre-industrial temperature change; (ii) compared to multi-decade period – 1960/70 – 1990; and (iii) compared to 1990. Given that pre-industrial temperatures are estimated to be 0.6°C lower than they are today, there is about a half a degree difference between options (i) and (ii) or (iii). Even this relatively 'small' difference could have significant implications for mitigation strategies. Unless otherwise noted, this paper refers the base period of 1990, which is consistent with much of the impacts literature cited here and reviewed in the Hitz and Smith paper (2004) and with the IPCC's treatment of this issue in the Third Assessment Report.

8   GMT increase serves as a proxy to discuss climate change impacts that occur at particular levels of atmospheric carbon concentrations (parts per million), but often this is a simplification. GMT also depends on factors other than carbon concentrations including: sulphate levels and the magnitude of their (negative) forcing, the climate sensitivity of the Earth in response to greenhouse gas build-up, the concentration of non-carbon greenhouse gases (e.g. methane, HFC, PFC and SF_6), and carbon uptake by oceans and other sinks. Because of these other factors, the strong positive correlation between atmospheric carbon concentrations and global mean temperature may or may not be linear. GMT is a good proxy for Article 2 discussions because it includes these other factors of climate change. However, for a given increase in GMT, there can be vast regional disparities in temperature, precipitation patterns, and climate variability. However, the more 'disaggregated' the climate indicators

(i.e. going from global to regional changes in climate), the more difficult it is to relate changes in GHG concentrations to climate outcomes.

9   It should be noted that broadening of sensitivity ranges, e.g. from Andronova and Schlesinger (2001) and Stainworth et al. (2005), make this more challenging. The statement assumes that we have confidence in the range of potential outcomes.

10 See also Dessai and Hulme (2004), Mastrandrea and Schneider (2004) and Wigley (2004).

11 Note that Figures 3 and 4 are not directly comparable, as they were derived from different models and prior assumptions. The former represents stabilization of atmospheric CO_2 concentrations, and is based on Monte Carlo analysis and uniform distribution of key parameters; whereas the latter shows stabilization of GHG for two particular emission scenarios, based on observations and expert elicitation, and non-uniform probability distributions. They are shown here to demonstrate two different ways of using probabilistic assessment of climate change outcomes as related to possible long-term objectives.

12 Nevertheless, a range of other authors have also suggested focusing on key vulnerabilities or impacts to climate change to inform global policy decisions (Pittock et al., 2001; Jones, 2004; Mastrandrea and Schneider, 2004; Jacoby, 2004; Patwardhan et al., 2003; Hare and Meinshausen, 2004; ECF and PIK, 2004; ICCT, 2005; UK Defra, 2005).

13 See Dessai et al. (2004), Jones (2004, 2001) and Jaeger (1998).

14 Examples of stakeholder engagement in impact assessment include the UK Climate Impacts Programme (ongoing) and USGCRP (2000), although neither of these address linkages to mitigation strategies. In two recent scientific meetings, UK Defra (2005) and ECF and PIK (2004), scientific experts and other stakeholders have been asked to reflect on interpretations of 'dangerous' climate change either on a regional or global basis. Also, since 1996, the European Union has had political consensus on the need to limit climate change to 2°C above pre-industrial levels (see EU, 2004, 2005). Their deliberations have included a general review of available impact information as well as information on the costs of mitigation.

15 Looking back, Yamin and Depledge (2004) discuss this with respect to negotiations (and negotiators) on the Protocol and the Convention.

16 In addition, the carbon fertilization effect saturates at higher CO_2 concentrations. Beyond roughly 600–800 ppm, there is little or no additional benefit to adding CO_2. Moreover, the higher CO_2 levels result in greater climate change, placing more stress on plants and ecosystems (Rosenzweig and Hillel, 1998).

17 For example, Hare (2005) finds that a 2°C increase in GMT above pre-industrial levels may already lead to potentially large extinctions or ecosystem collapses. See also ACIA (2004) for an understanding of regional risks of global climate change in the Arctic. In addition, while we refer here to 2°C as 'low' levels of change, Azar and Rodhe (1997) have pointed out that this is roughly double the increase experienced in the previous millennium occurring in a single century.

18 This issue is addressed in Schneider (2004) as well as in Narain and Fisher (2000), Baranzini et al. (2003), Mastrandrea and Schneider (2001, 2004), Yohe (2003; based on Roughgarden and Schneider, 1999) and Azar and Lindgren (2003). Nordhaus and Boyer (2000) also include what they call 'catastrophic' change and a limited set of natural system amenities, though they ignore other non-market impacts including irreversible ecosystem effects; contrary to the other studies noted above, they conclude that these events make no difference to the timing of abatement (though it might raise the costs in the distant future).

19 More recently, LRTAP has extended its reach to persistent organic pollutants – but since this experience is quite recent, the focus here is on acid rain.

References

ACIA, 2004. Impacts of a Warming Arctic: Arctic Climate Impact Assessment. Cambridge University Press, Cambridge, UK.

Agrawala, S., 1999. Early science–policy interactions in climate change: lessons from the Advisory Group on Greenhouse Gases Global Environmental Change 9, 157–169.

Agrawala, S., Raksakulthai, V., van Aalst, M., Larsen, P., Smith, J. and Reynolds, J., 2003. Development and Climate change in Nepal: Focus on Water Resources and Hydropower.Com/ENV/EPOC/DCD/DAC(2003)1/FINAL, OECD, Paris, 64pp.

Allen, M., Raper, S., Mitchell, J., 2001. Uncertainty in the IPCC's third assessment report. Science 293, 430–433.

Alley, R.B., Marotzke, J., Nordhaus, W.D., Overpeck, J.T., Peteet, D.M., Pielke, R.A. Jr, Pierrehumbert, R.T., Rhines, P.B., Stocker, T.F., Talley, L.D., Wallace, J.M., 2003. Abrupt climate change. Science 299, 2005–2010.

Andronova, N.E., Schlesinger, M.E., 2001. Objective estimation of the probability distribution for climate sensitivity. Journal of Geophysical Research 106, 22605–22612.

Azar, C., Lindgren, K., 2003. Catastrophic events and stochastic cost–benefit analysis of climate change. Climatic Change 56, 245–255.

Azar, C. and Rodhe, H., 1997. Targets for Stabilization of Atmospheric CO_2. Science 276, 1819–1819.

Baranzini, A., Chesney, M., Morisset, J., 2003. The impact of possible climate catastrophes on global warming policy. Energy Policy 31, 691–701.

Berk, M., van Minnen, J.G., Metz, B., Moomaw, W., den Elzen, M., van Vuuren, D., Gupta, J., 2002. Climate Options for the Long-Term (COOL). Global Dialogue – Synthesis Report, Report 410 200 118, RIVM, Bilthoven.

Cannon, T., 2002. Food security, development and climate change. Paper presented to the OECD Informal Expert Meeting on Development and Climate Change, 13–14 March 2002, Paris [available at http://www.oecd.org/env/cc].

Corfee Morlot, J., Agrawala, S., 2004. The benefits of climate policy. Global Environmental Change [Special Edition on the Benefits of Climate Policy] 14, 197–199.

Corfee Morlot, J., Höhne, N., 2003. Climate change: long-term targets and short-term commitments. Global Environmental Change 13, 277–293.

Depledge, J., 2000. Tracing the origins of the Kyoto Protocol: an article-by-article textual history. Technical Paper, Document No. FCCC/TP/2000/2 [available at http://www.unfccc.int].

Dessai, S., Adger, W.N., Hulme, M., Turnpenny, J., Köhler, J., Warren, R., 2004. Defining and experiencing dangerous climate change. Climatic Change 64(1–2), 11–26.

Dessai, S., Hulme, M., 2004. Does climate adaptation policy need probabilities? Climate Policy 4, 107–128.

ECF and PIK, 2004. What is dangerous climate change? Initial results of a symposium on Key Vulnerable Regions, Climate Change and Article 2 of the UNFCCC, European Climate Forum and Postdam Institute for Climate Impact Research [available at http://www.european-climate-forum.net/pdf/ECF_beijing_results.pdf].

Etterson, J.R., Shaw, R.G., 2001. Constraint to adaptive evolution in response to global warming. Science 294, 151–154.

EU, 2004. Council of the EU, Information Note from General Secretariat to Delegations, 22 December 2004: Climate Change: Medium and Longer Term Emission Reduction Strategies, Including targets. Council (Environment) Conclusions. Doc no: 16298/04/ENV 711/ENER 274/FISC 262/ONU 120 [also press release, available at http://eu.eu.int/ueDocs/cms_Data/docs/pressData/en/envir/83237.pdf].

EU, 2005. Council of the EU, Information Note from General Secretariat to Delegations, 11 March 2005: Climate Change: Medium and Longer Term Emission Reduction Strategies, Including Targets. Council (Environment) Conclusions. Doc No.: 7242/05/ENV 118/ENER 42/FISC 33/ONU 34.

Fischer, G., Shah, M. and van Velthuizen, H., 2002. Climate Change and Agricultural Vulnerability. IA–02–001. Jointly published by IIASA and FAO, Laxenburg.

Gitay, H., Brown, S., Easterling, W., Jallow, B. et al., 2001. Ecosystems and their goods and services. Climate change 2001: Impacts, Adaptation and Vulnerability, A Report of the Working Group II of the Intergovernmental panel on Climate Change, edited by J.J. McCarthy, O.F. Canziani, N.A. Leary, D.J. Dokken and K.S. White. Cambridge: Cambridge University Press pp. 331–342.

Hare, B., 2005. Relationship between increases in global mean temperature and impacts on ecosystems, food production, water and socio-economic systems. Paper presented at UK Defra scientific symposium on Stabilisation of Greenhouse Gases, 1–3 February, 2005, Exeter, UK.

Hare, B., Meinshausen, M., 2004. How much warming are we committed to and how much can be avoided? PIK Report, No. 93, Potsdam Institute for Climate Impact Research [available at http://www.pik-potsdam.de/publications/pik_reports].

Hitz, S., Smith, J., 2004. Estimating global impacts from climate change. In: The Benefits of Climate Policy: Improving Information for Policymakers. OECD, Paris.

ICCT, 2005. Meeting the Climate Challenge: Recommendations of the International Climate Change Taskforce. London, Washington, Canberra: The Institute for Public Policy Research, The Center for American Progress, The Australia Institute; January 2005.

IPCC, 2001a. Climate Change 2001. Synthesis Report. Cambridge University Press, Cambridge, UK.

IPCC, 2001b. Climate Change 2001: The Scientific Basis, A Report of the Working Group I of the Intergovernmental Panel on Climate Change, edited by J.T. Houghton, Y. Ding, D.J. Griggs, M. Noguer, P.J. van der Linden, X. Dai, K. Maskell and C.A. Johnson. Cambridge University Press, Cambridge, UK.

IPCC, 2001c. Climate Change 2001: Impacts, Adaptation, and Vulnerability, A Report of the Working Group II of the Intergovernmental Panel on Climate Change, edited by J.J. McCarthy, O.F. Canziani, N.A. Leary, D.J. Dokken and K.S. White. Cambridge University Press, Cambridge, UK.

IPCC, 2001d. Climate Change 2001: Mitigation, edited by B. Metz, O. Davidson, R. Swart and J. Pan. Cambridge University Press, Cambridge, UK.

Jacoby, H., 2004. Toward a framework for climate benefits estimation. In: The Benefits of Climate Policy: Improving Information for Policymakers. OECD, Paris.

Jaeger, C.C., 1998. Risk management and integrated assessment. Environmental Modelling and Assessment 3, 211–225.

Jones, R.N., 2001. An environmental risk assessment/management framework for climate change impact. Natural Hazards 23, 197–230.

Jones, R.N., 2004. Managing climate change risks. In: The Benefits of Climate Policy: Improving Information for Policymakers. OECD, Paris.

Klein, R.J.T., 2001. Adaptation to Climate Change in German Official Development Assistance: An Inventory of Activities and Opportunities, with a Special Focus on Africa. Deutsche Gesellschaft für Technische Zusammenarbeit, Eschborn, Germany.

Leemans, R., Eickhout, B., 2004. Another reason for concern: regional and global impacts on ecosystems for different levels of climate change. Global Environmental Change [Special Edition on the Benefits of Climate Policy] 14, 219–228.

Levy, M.A., 1993. European acid rain: the power of tote-board diplomacy. In: Haas, P.M., Keohane, R.O., Levy, M.A. (Eds), Institutions for the Earth: Sources of Effective International Environmental Protection. MIT Press, Cambridge, MA, pp. 75–132.

Mastrandrea, M., Schneider, S. 2001. Integrated assessment of abrupt climatic changes. Climate Policy 1, 433–449.

Mastrandrea, M., Schneider, S., 2004. Probabilistic integrated assessment of 'dangerous' climate change. Science 304, 571–575.

Metz, B., Berk, M., den Elzen, M., de Vries, B., van Vuuren, D., 2002. Towards an equitable global climate change regime: compatibility with Article 2 of the climate change convention and the link with sustainable development. Climate Policy 2(2–3), 211–230.

Narain, U., Fisher, A., 2000. Irreversibility, uncertainly, and catastrophic global warming. Giannini Foundation Working Paper 843. University of California, Department of Agriculture and Resource Economics, Berkeley, CA.

Nordhaus, W.D., Boyer, J.G., 2000. Warming the World: Economic Models of Global Warming . MIT Press, Cambridge, MA.

OECD, 2003a. Development and climate change project in Bangladesh: focus on coastal flooding and the Sundarbans. Document COM/ENV/EPOC/DCD/DAC(2003)3/FINAL.

OECD, 2003b. Development and climate change project in Fiji: focus on coastal mangroves. Document COM/ENV/EPOC/DCD/DAC(2003)4/FINAL.

OECD, 2003c. Development and climate change project in Nepal: focus on water resources and hydropower. Document COM/ENV/EPOC/DCD/DAC(2003)1/FINAL.

OECD, 2003d. Development and climate change project in Tanzania: focus on Mount Kilimanjaro. Document COM/ENV/EPOC/DCD/DAC(2003)5/FINAL.

OECD, 2004a. Development and climate change project in Egypt: focus on coastal resources and the Nile. Document COM/ENV/EPOC/DCD/DAC(2004)1/FINAL.

OECD, 2004b. Development and climate change project in Uruguay: focus on coastal zones, agriculture and forestry. Document COM/ENV/EPOC/DCD/DAC(2004)2/FINAL.

OECD, 2004c. The Benefits of Climate Policy: Improving Information for Policymakers, OECD, Paris

O'Neill, B.C., Oppenheimer, M., 2002. Dangerous climate impacts and the Kyoto Protocol. Science 296, 1971–1972.

O'Neill, B.C., Oppenheimer, M., 2004. Climate change impacts sensitive to path to stabilization. Proceedings of the National Academy of Science 101(16), 411–416.

Oppenheimer, M., Alley, R.B., 2004. The West Antarctic ice sheet and long term climate policy. Climatic Change 64, 1–10.

Oppenheimer, M., Alley, R.B., 2005. Ice sheets, global warming, and Article 2 of the UNFCCC. Climatic Change 68: 257–269.

Oppenheimer, M., Petsonk, A., 2005. Article 2 of the UNFCCC: historical origins, recent interpretations. Climatic Change, in press.

Parmesan, C., Yohe, G., 2003. A globally coherent fingerprint of climate change impacts across natural systems. Nature 421, 37–41.

Parry, M.L., Hossell, J.E., Jones, P.J., Rehman, T., Tranter, R.B., Marsh, J.S., Rosenzweig, C., Fischer, G., Carson, L.G., Bunce, R.G.H., 1996. Integrating global and regional analyses of the effects of climate change: a case study of land use in England and Wales. Climatic Change 32(2), 185–198.

Patwardhan, A., Schneider, S.H., Semenov, S.M., 2003. Assessing the science to address UNFCCC Article 2: a concept paper relating to cross cutting theme number four (IPCC Concept Paper) [available at http://www.ipcc.ch/activity/cct3.pdf].

Pershing, J., Tudela, F., 2003. A long-term target: framing the climate effort. In: Beyond Kyoto: Advancing the International Effort Against Climate Change, December 2003. Pew Center on Global Climate Change, Washington, DC, pp. 11–36.

Pingali, P., 2004. Key sensitivities and levels of impact in selected areas and regions. Paper presented to IPCC Expert Meeting on The Science to Address UNFCCC Article 2 including Key Vulnerabilities, Buenos Aires, Argentina, 18–20 May 2004, meeting report [available at http://www.ipcc-wg2.org].

Pittock, A.B., Jones, R.N., Mitchell, C.D., 2001. Probabilities will help us plan for climate change. Nature 413, 249.

Reilly, J., Stone, P., Forest, C., Webster, M., Jacoby, H., Prinn, R., 2001. Uncertainty and climate change assessments, Science 293, 430–433.

Richardson, S.D., Reynolds, J.M., 2000. An overview of glacial hazards in the Himalayas. Quaternary International 65/66, 31–47.

Root, T.L., Price, J.T., Hall, K.R., Schneider, S., Rosenzweig, C., Pounds, A., 2003. Fingerprints of global warming on wild animals and plants. Nature 421, 57–60.

Rosenzweig, C., Hillel, D., 1998. Climate Change and the Global Harvest: Potential Impacts of the Greenhouse Effect on Agriculture. Oxford University Press, New York.

Roughgarden, T., Schneider, S.H., 1999. Climate change policy: quantifying uncertainties for damages and optimal carbon taxes. Energy Policy 27(7), 415–429.

Schellnhuber, J., Warren, R., Haxeltine, A., Naylor, L., 2004. Integrated assessment of benefits of climate policy. In: The Benefits of Climate Policy: Improving Information for Policymakers. OECD, Paris.

Schneider, S.H., 2004. Abrupt non-linear climate change, irreversibility and surprise. Global Environmental Change 14: 245–258.

Schneider, S.H., Lane, J., 2004. Abrupt non-linear climate change and climate policy. In: The Benefits of Climate Policy: Improving Information for Policymakers. OECD, Paris.

Schwartz, P., Randall, D., 2003. An abrupt climate change scenario and its implications for United States national security. A report commissioned by the US Defense Department [available at http://www.gbn.com/GBNDocumentDisplayServlet. srv?aid=26231&url=%2FUploadDocumentDisplayServlet.srv%3Fid%3D28566].

Semb, A., 2002. Sulphur dioxide: from protection of human lungs to remote lake restoration. In: Harramoës, P., Gee, D., MacGarvin, M., Stirling, A., Keys, J., Wynne, B., Guedes Vaz, S. (Eds), The Precautionary Principle in the 20th Century. Earthscan, London.

Shrestha, M.L. and Shrestha, A.B., 2004. Recent Trends and Potential Climate Change Impacts on Glacier Retreat/Glacier Lakes in Nepal and Potential Adaptation Measures. Paper presented at the OECD Global Forum on Sustainable Development: Development and Climate Change, ENV/EPOC/GF/SD/RD(2004)6/FINAL, OECD, Paris, 64pp.

Smith, J.B., Schellnhuber, H.-J., Mirza, M.Q., 2001. Lines of evidence for vulnerability to climate change: a synthesis. In: McCarthy, J.J., Canziani, O.F., Leary, N.A., Dokken, D.J., White, K.S. (Eds), Climate Change 2001: Impacts, Adaptation, and Vulnerability, A Report of the Working Group II of the Intergovernmental Panel on Climate Change. Cambridge University Press, Cambridge, UK, pp. 914–967.

Stainworth, D.A., Aina, T., Christensen, C., Collins, M., Fauli, N., Frame, D.J., Kettleborough, J.A., Knight, S., Martin, A., Murphy, J.M., Piani, C., Sexton, D., Smith, L.A., Spicer, R.A., Thorpe, A.J., Allen, M.R., 2005. Uncertainty in predictions of the climate response to rising levels of greenhouse gases. Nature 433, 403–406.

Swart, R.J., Vellinga, P., 1994. The 'ultimate objective' of the Framework Convention on Climate Change requires a new approach in climate change research. Climatic Change 26(4), 343–350.

Thomas, C.D., Cameron, A., Green, R.E., Bakkenes, M., Beaumont, L.J., Collingham, Y.C., Erasmus, B.F.N., Ferreira de Siqueira, M., Grainger, A., Hannah, L., Hughes, L., Huntley, B., van Jaarsveld, A.S., Midgley, G.F., Miles, L., Ortega-Hueta, M.A., Perterson, A.T., Phillips, O.L., Williams, S.E., 2004. Extinction risk from climate change. Nature 427, 145–148.

UK Defra, 2005. Report on Avoiding Dangerous Climate Change: A Scientific Symposium on Stabilisation of Greenhouse Gases, 1–3 February, 2005, Exeter, UK [available at http://www.stabilisation2005.com].

UNFCCC, 1992. United Nations Framework Convention on Climate Change, International Legal Materials 31, 1992, 849 [available at http://www.unfccc.int].

USGCRP, 2000. National Assessment Synthesis Team, Climate Change Impacts on the United States: The Potential Consequences of Climate Variability and Change. US Global Change Research Program, Washington, DC.

Webster, M., Forest, C., Reilly, J., Babiker, M., Kicklighter, D., Mayer, M., Prinn, R., Sarofim, M., Sokolov, A., Stone, P., Wang, C., 2003. Uncertainty analysis of climate change and policy response. Climatic Change 63, 295–320.

Wigley, T.M.L., 2004a. Modelling climate change under no-policy and policy emission pathways. In: The Benefits of Climate Policy: Improving Information for Policymakers. OECD, Paris.

Wigley, T.M.L., 2004b. Choosing a stabilization target for CO_2. Climatic Change 67, 1–11.

Wigley, T.M.L., Raper, S.C.B., 2001. Interpretations of high projections for global mean warming, Science 293, 451–454.

Yamin, F., Depledge, J., 2004. The International Climate Change Regime: A Guide to Rules, Institutions and Procedures. Cambridge University Press, Cambridge, UK.

Yohe, G., 2003. Estimating benefits: other issues concerning market impacts. OECD, Paris (ENV/EPOC/GSP(2003)8/FINAL).

Climate Policy 5 (2005) 273–290

Between the USA and the South: strategic choices for European climate policy

Frank Biermann*

Department of Environmental Policy Analysis, Institute for Environmental Studies (IVM), Vrije Universiteit Amsterdam, De Boelelaan 1087, 1081 HV Amsterdam, The Netherlands

Received 14 January 2005; received in revised form 8 March 2005; accepted 8 March 2005

Abstract

This article discusses Europe's strategic choices in current climate policy. It argues that the future climate governance architecture must pass four tests: credibility, stability, flexibility, and inclusiveness. Drawing on this, I review the strategic choices for Europe, structured around three levels of analysis in political science: climate polity, that is, the larger institutional and legal context of policy making; climate policy, the instruments and targets that governments agree to implement; and climate politics, the actual negotiation process. At each level of analysis, I look at the interests and expectations of two non-European actors or actor groups: the USA, which accounts for over a third of all Northern greenhouse gas emissions, and the group of developing countries, which accounts for the vast majority of humankind. I argue that Europe must take clear principled positions on a number of key issues, in particular the need to have a strong multilateral framework as the sole and core institutional setting for climate policy and to accept the principle of equal per-capita emissions entitlements as the long-term normative bedrock of global climate governance. Both positions, however, will alienate the USA, and both will make it more difficult for the USA to rejoin the international community on the climate issue.

Keywords: European climate policy; US climate policy; Developing country climate policy; Post-2012 climate governance architecture; Kyoto Protocol

Introduction

Europe's position in climate policy is unique. Europe is one of the regions most concerned about climate change, yet it ranks among the highest per-capita carbon dioxide emitters in the world. Europe claims to speak as one actor on the global stage, but suffers from the need for tedious intra-European negotiation and coordination. Europe wishes to take the lead in global climate policy, but struggles with the consequences of this first major attempt at multilateral regime creation without the consent and support of the USA, as well as with its own difficulties in meeting the Kyoto targets. Europeans also like to see themselves as the more cooperative part of the industrialized North when it comes to multilateralism, support for international institutions and

* Corresponding author. Tel.: +31-20-59-89959; fax: +31-20-59-89553
E-mail address: frank.biermann@ivm.vu.nl

organizations, and collaboration with developing countries in fighting poverty: but at the same time, European contributions to bilateral and multilateral development assistance programmes have dwindled over the last decades.

Yet Europe's unique position makes it also an ideal bridge between conflicting interests in climate governance (and beyond), notably between the anti-Kyoto coalition in the North – now reduced to the USA and Australia – and the developing world with its plethora of interest constellations. This bridge-position of Europe between the rest of the North and the global South is at the centre of this article. What strategic choices does Europe have when it comes to climate governance in the 21st century? What kind of institutions and governance systems will guarantee a safe landing of the global climate system? What are the different expectations and interests from outside Europe that European policymakers are faced with?

Since the structure of any effective governance system must be tailored to the specific problem at hand,[1] I start with the premise that in climate policy, institution-building will have to convincingly deal with five characteristics:

1. Persistent uncertainties regarding causes, impacts, interlinkages, and nonlinearity;
2. Long, potentially irreversible cause-and-effect relationships and hence planning horizons that surpass the tenure, and even the lifetime, of most present decision-makers and stakeholders;
3. Complex interlinkages between different policy areas within and beyond climate policy;
4. Global interdependence and mutual substitutability of response options (for every global policy target, there is an unlimited number of possible combinations of local responses across nations and time frames with equal degrees of effectiveness);
5. Possibly devastating climate change impacts that current governance systems might not be fully prepared for.

This problem-structure is unprecedented in international relations. Transnational institutions with a time-horizon of centuries are rare – the Catholic Church, with its 2,000-year stable leadership succession and decision-making mechanisms is probably the only empirical example. Stratospheric ozone depletion shares some characteristics with the climate problem, but is less problematic given the availability of substitutes, the restricted use of harmful substances, and the relative confidence of the underlying science. Nuclear proliferation shares a few characteristics, such as global interdependence and catastrophic threats, but also displays much dissimilarity.

Given this complex problem-structure, I argue, as a further, derived premise, that the future climate governance architecture must pass four tests:

1. *Credibility*: Some governments will have to commit resources both domestically and through transnational transfer mechanisms towards solving this problem, based on the assumption that other governments will reciprocate when it is their turn (including governments to come in the future). The climate governance system must thus produce the necessary credibility for governments to believe in this reciprocity over time.
2. *Stability*: This requires that the future climate governance system must be stable enough to withstand political changes in participating countries and altered international interest coalitions.
3. *Flexibility*: Within this stable framework, future governments must have the ability, based on previously agreed procedures and principles, to change regime elements to respond to new

situations and new scientif ic findings, without harming the credibility and stability of the entire system.

4. *Inclusiveness*: The interdependence of current climate politics, as well as the complexity and uncertainty of the entire climate system that may change the overall interest constellation within a few years, require the governance system to be as inclusive as possible regarding the number of parties involved and the emissions represented by them.

One cross-cutting requirement of these four governance characteristics is the need to establish universally accepted basic norms and problem frames among states. Recent research in the field of international relations[2] has indicated that the political behaviour of states cannot be explained merely through simple calculations of material interest and power, as earlier theories in the framework of political realism had posited. Instead, states are guided in their behaviour by international norms that prescribe and prohibit certain types of behaviour and that create an international society that 'socializes' states – including new governments that have not participated in the original creation of norms, as will be the case for future rounds of climate policy in the decades to come. To be effective, norms must be relatively simple, they must be cross-culturally appealing, and they must be sufficiently clear and unambiguous. For example, the success of the world trade regime in liberalizing trade and phasing out most custom duties within half a century is partially attributed to the simplicity and general acceptability of its basic principles, notably reciprocity and the most-favoured-nation clause. A further example is the development of human rights norms in the course of the 20th century (Risse et al., 1999). In the climate regime, it appears that in particular the definition of Article 2 of the UN Framework Convention on Climate Change, as well as the basic allocation criteria for greenhouse gas emissions, will require the establishment of global norms that have a high degree of simplicity, acceptability, and unambiguity. A second cross-cutting requirement is the need to establish enforceable norms (though not necessarily a regime with strong sanctions). Enforceability means that governments must be able to assess both the normatively required and the factually implemented contribution of other nations to the solution of the problem, once significant costs are involved. This relates both to the clarity of the commitments and the availability of monitoring and reporting mechanisms.

Drawing on this discussion, I now review in detail the strategic choices for Europe in the global context.[3] The text is structured around the three key levels of analysis in political science: climate *polity*, that is, the larger institutional and legal context of policy making; climate *policy*, the actual instruments and targets that governments agree to implement; and climate *politics*, the actual negotiation process. At each of these levels, I look at the interests and expectations of two non-European actors or actor groups:[4] the USA, which accounts for over one-third of all Northern emissions, and the group of developing countries, which accounts for the vast majority of humankind. The actor-quality of the latter has been in doubt since the 1970s, yet has withstood all doomsayers through the continuous, although weakened, existence of the 'Group of 77/China', which now includes 132 developing nations. Internal differentiation processes have resulted in conflicting interests within the Group of 77 in many policy domains, such as the negotiations on the law of the sea or trade in agricultural products – and climate policy is yet another area that reveals different interests and strategies within the group. And yet, when larger issues of North–South confrontation are at stake, the Group of 77 still demonstrates remarkable unity (given the large number of countries and interests involved), which expresses itself in the essentially uniform treatment of developing countries in both the climate convention and its Kyoto Protocol. However, given existing internal conflicts, I focus largely on the 'mainstream' of the major, most powerful, developing countries, while neglecting special groupings such as oil-producing nations or small island states.[5]

Climate polity

In the upcoming negotiations of the post-2012 climate regime, the European Union will have to develop a clear vision, not only about the most appropriate climate policies, such as short-term targets and timetables, but also the best climate *polity* – the institutional setting and governance system in which future policies are negotiated, monitored, implemented and internationally enforced. Europe, however, is faced with conflicting demands from different countries.

The US perspective

On the one hand, the USA, Europe's main traditional partner, has formally rejected the Kyoto Protocol, which the Bush administration has determined to be 'fatally flawed'.[6] In addition, the US Senate, which has to ratify any international agreement, determined in 1997 in a unanimous decision sponsored by senators Byrd and Hagel with 65 bipartisan co-sponsors,[7] that 'the exemption for Developing Country Parties is inconsistent with the need for global action on climate change and is environmentally flawed' and that 'the proposals under negotiation ... could result in serious harm to the United States economy, including significant job loss, trade disadvantages, increased energy and consumer costs, or any combination thereof'.[8] This perception in the USA has hardly changed. According to Senator Byrd, 5 years later, 'the conditions outlined in that [Byrd–Hagel] resolution remain the guideposts for U.S. international climate change policy'.[9] In 2003, Paula Dobriansky, the US Under-Secretary of State for Global Affairs, informed delegates at the conference of the parties to the climate convention of the US position that the Kyoto Protocol was 'an unrealistic and ever-increasing regulatory straitjacket'.[10] It is unlikely that the entry into force of the protocol on 16 February 2005 and its almost universal acceptance will encourage US participation, given the widespread resistance within the US government towards the protocol, including in the US Senate where a vote on the instrument of ratification would require a two-thirds majority.

The unilateral rejection of the Kyoto Protocol by the USA is no exception but fits into a larger pattern of unilateral policy making. The USA has hindered progress or rejected a number of widely accepted core projects of global governance, including the International Criminal Court, the international treaty on the prohibition of anti-personnel mines, and even the convention on the rights of the child and the convention on the elimination of all forms of discrimination against women. Also many environmental treaties function without the USA, including the biodiversity convention of 1992 and its Cartagena Protocol on safety in the trade of genetically modified organisms, the Basel agreement on the transboundary shipment of hazardous waste and its disposal, and the recent conventions on prior informed consent procedure for certain hazardous chemicals and on persistent organic pollutants.[11] The USA's rejection of the Kyoto Protocol, despite it being almost universally accepted with ratifications by 141 countries, is hence merely one example of a larger pattern in US foreign policy.

Given this situation, a number of (mostly) US researchers have developed a stream of proposals of alternatives to the Kyoto architecture that could allow the USA to (re-)engage in international efforts. First, US actors have tried to encourage developing countries to turn away from the Kyoto basic framework and to adopt some form of commitments, including voluntary targets of some kind, within or outside the Kyoto system. This strategy has so far only been successful with Argentina, which in 1998 adopted voluntary targets for greenhouse gas emissions (partially motivated by its attempt to join the OECD), and with Kazakhstan, which is not a traditional, mainstream developing country (see details in Egenhofer and Fujiwara, 2003). This US coaxing is often motivated by a popular US problem perception that is framed in aggregate emission data on country level, without consideration of population size or the level of development – such as the rhetoric that 'China will surpass America in greenhouse gas emissions by 2015'.[12] Given the current US debate on 'Kyoto', however, it is doubtful whether the USA will ratify the protocol even in the unlikely event that some developing countries agree to quantified emissions limitation commitments.

Second, some US authors have suggested that the USA should conclude alternative, regional agreements with like-minded countries, for example in Latin America[13] or with China and possibly other key developing countries (Stewart and Wiener, 2003). Daniel Bodansky, for instance, argues for an 'institutional hedging strategy' with the USA becoming the creator of 'a more diversified, robust portfolio of international climate change policies in the long term' (Bodansky, 2002a, p. 1). Such regional or small-party agreements could cover only the world's largest greenhouse gas emitters and would allow for experimentation with alternative international climate regulatory frameworks. For some, such an approach would allow negotiation with only the more 'moderate' developing countries, while disabling 'the hard-line developing countries [...] to prevent more moderate developing states from joining' (Bodansky, 2002a, p. 6). At some point, however, these regional regimes under US leadership could again allow for reintegration of the world into a single global regime (Sugiyama, 2003; Bodansky, 2002a).

Third, and related to the second point, US authors are often less than enthusiastic about the role of the United Nations in climate governance, and see the UN system as part of the problem rather than a solution. Negotiations under the UN umbrella are seen as being too 'large, unwieldy, ideologically laden' to oversee the 'simple tasks of the kind required [under the climate convention]' (Bodansky, 2002a, 3) – an argument starkly in contrast with the Southern view that refuses to accept climate governance as a 'simple', technical issue and that supports the UN as a body where numbers count and Southern interests are respected.

Fourth, US authors have proposed joint action through issue-specific agreements (cf. the analysis in van Asselt et al., 2004). These could include targets for specific sectors, e.g. energy-efficiency standards for the global automobile industry that would need to bring together only the most important car-producing countries (Barrett, 2002, p. 6); specific policies, such as energy taxes; agreements and targets for cooperation in scientific research and technology development (Benedick, 2001; Aldy et al., 2003; Barrett, 2002; Tol, 2002), including on carbon sequestration[14], renewables, geological storage and energy conservation; targets on technology transfer and capacity building; and specific measures to increase cooperation regarding adaptation.

Fifth, the USA is at the forefront of promoting public–private and private–private partnerships at the global level, notably in supporting the so-called 'type-2 outcomes' of the 2002 World Summit on Sustainable Development (as opposed to type-1 traditional intergovernmental agreements) (for an overview, see Hale and Mauzerall, 2004).

Not all of these proposals and approaches are exclusively linked to US authors or to the US administration,[15] nor do they all deny a role for, or the continuing existence of, a multilateral framework. In theory, issue-specific or regional agreements could be reached outside the climate convention, but also within its framework (for example in the form of new protocols), or even as part of an amended Kyoto Protocol. Many proposals also either suggest a role for the climate convention as facilitator of other approaches (Sugiyama, 2003) or envisage a return to this framework over time (e.g. Stewart and Wiener, 2003; Bodansky, 2002a, 2002b). Likewise, the US administration does not represent 'all America': many promising initiatives on climate policy are under way in the USA below the level of federal policy, including at state level[16] or through private agreements (including the Chicago Climate Exchange, created in 2001). Yet, in sum, the US administration and large sections of its academic and policy communities expect Europe to be open to international agreements of a different kind, and implicitly to abandon the pursuance of the Kyoto accord and the 'Kyoto approach' that the Bush administration has determined (like quite a few other agreements) to be 'flawed'.

The Southern perspective

Developing countries, on their part, continue to support the multilateral approach in climate policy as in other policy domains, and most Southern nations have ratified the Kyoto Protocol. Multilateralism allows the South to count on its numbers in diplomatic conferences and gain bargaining power from a uniform negotiation position; it allows for side-payments across negotiation clusters within a policy domain and across different policies; and it minimizes the risk of developing countries being coerced into bilateral agreements with powerful nations that might offer them suboptimal negotiation outcomes. For the many smaller and medium-sized developing countries, unity is strength, and multilateralism is its only guarantee. Since the emergence of the climate issue, the South has therefore sought to bring all negotiations under the UN framework and to frame global warming as an overarching political problem with implications far beyond mere environmental policy. Therefore, climate change was located directly under the UN General Assembly, which declared the issue in 1988 a 'common concern of [hu]mankind'[17] and convened the intergovernmental negotiation process two years later. Climate change was hence framed as a prime example of a problem of environment *and* development, and it remained in the negotiations outside the coordinating influence of the issue-specific UN Environment Programme.

Europe's strategic choice

The European Union stands between conflicting expectations: joining the USA in its attempt to scuttle the multilateral process and to engage in sectoral or selective agreements between a few like-minded partners, or supporting the South in its wish to keep all institutional and policy development under the overall umbrella of the UN and of the UN climate agreements. Both strategies can coexist only to a limited extent: the European Union needs to make strategic choices.

On the one hand, sectoral, selective or non-legally-binding agreements bear the promise of quicker solutions, since negotiations are easier given the smaller number of actors and interests at the table and the non-binding nature of the outcome. The advantage of fewer parties has been emphasized by some strands of negotiation theory, which posit stronger commitments and faster progress the fewer (like-minded) parties participate in a given negotiation. Some cite the 1987 Montreal Protocol on Substances that Deplete the Ozone Layer (e.g. Simpson, 2002, p. 72), which was negotiated in the mid-1980s within a small group of industrialized countries with only a few developing countries involved.

On the other hand, a selective agreement will produce, first, a solution that applies only to the few participating countries and fits the interests only of those countries, and it is not guaranteed that other countries will later join. Second, a quick success in negotiating sectoral agreements might run counter to long-term success, when important structural regime elements have not been sufficiently resolved. Third, smaller agreements with only a few 'like-minded' countries will decrease the opportunity for creating package deals, which will minimize the overall policy acceptance and effectiveness (Sugiyama, 2003; Tangen and Hasselknippe, 2003). The 1987 Montreal Protocol illustrates all three problems: Even though the protocol was relatively quickly negotiated within the OECD group, it was subsequently not accepted by the major developing countries. Two years after adoption of the protocol, only 10 had ratified the treaty, and of the 13 developing countries whose CFC consumption appeared to rise most sharply in 1987, only Mexico, Nigeria and Venezuela had joined (Kohler et al., 1987). In August 1989, a UN working group[18] therefore warned that 'for the Protocol to be fully effective in its purpose of controlling the emissions of CFCs and halons, all countries must become Parties'. Both China and India agreed to ratify the treaty only after substantial changes to its basic structure had been made.

In the ozone regime, the Southern contribution to the problem was small, yet threatened to grow. In climate governance, the Southern role is much larger from the outset. Regional agreements of a few like-minded players, in the hope that others will later follow, do not promise to bring the long-term trust and regime stability that is needed in the climate domain. An 'institutional hedging strategy' (Bodansky, 2002a) with different policies and regimes scattered around the globe might seem to be a novel and reasonable solution at first sight. In the long run, however, such a strategy would cause havoc to the larger goal of building a truly universal climate governance architecture. The post-2012 climate governance system requires institutional mechanisms that are trustworthy, stable, flexible, provide for cross-issue bargains, and include all nations. This can be offered only by a global framework agreement that sets out the constitutional rules of climate governance in the 21st century, and detailed agreements on sub-questions that are negotiated within the larger, stable, normative system that sustains the 'grand bargain' (see also Müller et al., 2003). Europe needs to play a major role in this centennial project.

Climate policy

A second key strategic decision for the European Union is the choice of the 'right' policy to be supported. To the extent that the Kyoto approach will continue to be followed, governments soon need to agree on new sets of targets and commitments for different countries after 2012. Key decisions to be taken refer to the kinds of commitments (degree of binding force), their time frame, and their content (degree and form of differentiation). Again, Europe is faced with varying demands from different parts of the world.

Type and time frame of commitments

The US perspective

In the USA, the Kyoto Protocol is largely perceived as asking too much too soon from its parties. US authors propose instead that targets should be lowered or abandoned in the short term (but may be stricter in the long run) (Benedick, 2001; Barrett, 2001, 2002); should have a variety of safety valves, such as price caps in the case of emissions trading (Aldy et al., 2001; Bodansky, 2002a; Victor, 2001) or an 'uncapped' emissions trading (Bradford, 2002); should be replaced (or strongly complemented) by market-based mechanisms (Bradford, 2002), by transnationally harmonized environmental taxation (Cooper, 1998, 2001), or by sectoral, issue-specific agreements (as outlined above) (e.g., Barrett, 2001, 2002; Benedick, 2001); and should guarantee or at least provide strong incentives and mechanisms to ensure that developing countries will also be covered at some time, for example through voluntary measures (Aldy et al., 2001; Stewart and Wiener, 2003).

The official US Global Climate Change Initiative of 2002 does not directly relate to the Kyoto Protocol – which is rejected – but includes a unilateral target for reducing the greenhouse gas intensity of the US economy by 18% until 2012 (which comes close to business-as-usual trends in the USA and will further increase total US emissions). The Initiative also plans to encourage industry to voluntarily adopt non-binding emissions intensity targets and to provide additional support for research and technology development, including carbon sequestration and hydrogen fuels.

The Southern perspective

Developing countries, on their part, expect industrialized countries to act; that is, to send a clear signal regarding their seriousness by committing themselves to enforceable and demanding short-term (and long-term) targets. Such targets are rejected for reduction commitments of the South itself (see also Simpson, 2002, pp. 45 et seq.). India's (then) Prime Minister, Atal Bihari Vajpayee, when opening the 2002 Conference of the Parties to the Climate Convention in New Delhi, listed all reasons usually given by Southern representatives when rejecting early targets for the South: First, that per capita greenhouse gas emissions are only a fraction of the world average and an order of magnitude below that of many developed countries; second, that 'the ethos of democracy' cannot support 'any norm other than equal per capita rights to global environmental resources'; third, that Southern per-capita incomes are a small fraction of those in industrialized countries; fourth, that developing countries lack adequate resources to meet their basic human needs and that climate change mitigation will bring additional strain to already fragile economies of the South; and fifth, that the greenhouse gas intensity of Southern economies at purchasing power parity is low and in any case below that of industrialized countries (Vajpayee, 2002).

The Delhi Declaration on Climate Change and Sustainable Development, adopted at the same conference, has again emphasized that development and poverty eradication are the overriding priorities of the South, and that nations have common, but clearly also differentiated, responsibilities. In the words of Thomas Schelling, 'there is no likelihood that China, India, Indonesia, Brazil or Nigeria will fully participate in any greenhouse-gas regime for the next few decades. They have done their best to make that point clear, and it serves no purpose to disbelieve them.'[19] Instead, developing countries highlight the core North–South compromise of the Climate Convention, notably Article 4.3, which obliges the North to provide new and additional financial resources to meet the agreed full incremental costs of measures agreed between developing countries and the Global Environment Facility, and Article 4.7, which stipulates that the extent to which developing countries will effectively implement their commitments under the convention will depend on the effective implementation by industrialized countries of their commitments under the convention related to financial resources and the transfer of technology (Biermann, 1999, 2002). Finally, many developing countries point to their real achievements in combating greenhouse gas emissions through a variety of measures, including through fiscal reform, economic restructuring, or renewable energy programmes (Müller et al., 2003).

Europe's strategic choice

What approach should the European Union follow? As for the South, it is unlikely that developing countries will accept quantified commitments in the near future; in the long run, all will depend on the degree of North–South differentiation that is being negotiated (discussed below).

Regarding the US perspective that argues against short-term targets, it seems unlikely from a political science perspective that exclusively long-term targets will do much good. The key issue in building the global climate governance system is trust and credibility. Nations that incur costs today need to be able to assume that other nations will do the same tomorrow. Long-term targets that do not build on short-term targets derived from long-term visions will lack such credibility. It will be comparatively easy to find agreement on targets for 2030 or 2050 if no clear policy commitment for the present is involved. Yet it will be difficult for decision-makers to believe that other nations will in fact adhere to such long-term targets, and it will be impossible to verify. No regime exists today that commits nations to targets 50 years from now. The world trade regime, the international financial and economic order,[20] and all environmental regimes commit parties to obligations right here and now. None allows nations to defer compliance to a time when all the present decision-makers are retired and no longer accountable.

Safety valves may increase participation in the short run, but reduce incentives for countries to enact strong domestic measures in the present. The same is true for regimes with low enforcement procedures. Informal, essentially non-binding, pledge-and-review and blame-and-shame systems have worked in the early stages of environmental regimes, in the process of broadening coalitions or in other policy areas, such as human rights. In issue areas where sizeable governmental expenses and societal losses are at stake, governments usually rely on clear, enforceable obligations often even subjected to independent adjudication. This is the success story of phasing out border customs through the General Agreement on Tariffs and Trade, and it also stood at the centre of the global phase-out of ozone-depleting substances. Clear, enforceable, short-term obligations are hence what the European Union needs to support in climate governance, too.

Degree of commitments and form of differentiation

The agreement on the time frame and type of future agreements will eventually depend on the degree of commitments to be negotiated, and here in particular on the allocation of commitments to each negotiation partner. The differentiation of obligations between nations under the Kyoto Protocol is based on political criteria, such as the exemption of all members of the political 'Group of 77' from quantified reduction commitments, and within the OECD block on a mix of economic costs and negotiation power and skills. This *ad hoc* procedure was inevitable in order to find sufficient support in the international community for creating the climate regime in the first place. Yet it is doubtful whether this willingness-to-pay approach can continue after 2012. At some point, generally agreed *a priori* criteria for the definition of a country's emissions reduction obligation will be required.

The US perspective

The USA and Australia seek to link future commitments to (a combination of) different principles, such as the principle of equal allocation of emissions rights to each unit of gross national product (which could be seen as a principle of need), the principle of protection of 'acquired' past emission rights (grandfathering), or the principle of equal energy efficiency (carbon-intensity targets). Again, these principles correlate and would result, if implemented, in comparably smaller obligations for richer industrialized countries. The sum of these principles could also be seen as the principle of equal cost of environmental policy, since each nation will have to reduce emissions in proportion to its overall economic activity. A sizable reallocation of economic resources between countries is unlikely under the principles submitted by the USA, which are hence socially conservative.

The Southern perspective

By and large, actors from the South strive for the early acceptance of one, some, or all of the following principles:

1. The principle of equal entitlement of all human beings to equal emissions (allocation of emission rights to countries based on their current population)
2. The principle of historic responsibility (allocation of current emission rights in negative correlation to the amount of past emissions)
3. The principle of basic or survival emissions (relief of countries from reduction obligations below a certain flat rate basic emission)
4. The principle of economic acceptability within the context of poverty reduction (relief from reduction commitments if the level of development is below certain levels) (for more detail on Southern perspectives, see Gupta, 1997; Najam et al., 2003; Simpson, 2002).

While these principles are derived from different claims to overarching principles of justice and fairness, they correlate and would, if implemented, exempt developing countries from obligations largely in proportion to their economic development and wealth.

Underlying this conflict between different principles is the overall conceptualization of the global warming problem, especially whether it is being framed as a global resource (to be allocated) or as an environmental problem (to be resolved). The Southern conception comes down to the view that the Earth's greenhouse gas absorption capacity constitutes a global resource to be allotted to humankind based on need (that is, in favour of the poorest) or based on equal per-capita entitlements (essentially a human-rights claim): this was the approach taken by the South also regarding the allotment of deep seabed mining resources, Antarctica, the geostationary orbit (here linked to quasi-territorial claims), or even 'the moon and other celestial bodies' (as a 1970s treaty read). Under-utilizing one's own quota of resources would then justify a transnational wealth transfer of those who wish to over-utilize their share. The US conceptualization, on its part, instead views the global warming problem as a burden to be fairly shared by all nations in a way that allows all partners, rich and poor, to suffer to comparable degrees and as little as possible. Transnational wealth transfer is, in such a view, unjustified,[21] and existing entitlements to wealth are protected and to be conserved.

Both sets of principles are mutually exclusive in their pure formula (though the US perspective could, of course, allow for exemptions for the poorest and most needy). However, all allocation principles can be combined in mixed formulas, and it is here that substantial debate has centred (for a discussion of options and further references, see Philibert and Pershing, 2001). A mixed formula – e.g. the allocation of emission entitlements based on weighted indicators for population and wealth, or population and grandfathering (such as contract-and-converge) – seems to be the only way to develop a sufficiently broad compromise for the upcoming years. This could be linked to triggering thresholds, such as the relief from commitments of all developing countries with per-capita emissions below the global average (which would serve as a safety valve for developing countries to enter global agreements) (see Müller et al., 2003, p. 2-7). A mixed formula could also be negotiated within a transition trajectory that eventually leads to one final basic formula, such as equal per-capita emission allotments ('contract and converge').

Europe's strategic choice

The European Union, again, stands between the different positions, and it appears likely that the Union will have to take the lead to foster a global compromise. A number of arguments speak in favour of supporting, as a long-term goal, the principle of equal per-capita emissions; that is, accepting mixed formulas only as trajectory to a final allocation based on one principle. First, there is some precedent in the Montreal Protocol, which used per-capita emissions in its definition of privileged developing countries in a system that for the most part worked well and was widely accepted, including by the USA.[22] Second, per-capita entitlements have an inherent appeal, due to their link to basic notions of human rights, which attracts populations in both the South and the North – the European philosopher Kant (1795/1983) affirmed 200 years ago the 'Recht des gemeinschaftlichen Besitzes der Oberfläche der Erde' [right of common ownership of the Earth's surface]. Third, only such a clear principle is likely to have the normative power to grant the climate governance system the institutional stability it needs in the decades and centuries to come – a complicated multifactor formula will not only be unlikely to bring stability, strength and clear expectations, but will also have to be continuously renegotiated, which will harm climate governance in the long run. Fourth and more practically, it seems unlikely that the developing world will ever sign up to any serious climate change agreement that includes the South but does not build on the final principle of equal per-capita entitlements. In the unequivocal words of the Indian Prime Minister at the 2002 Conference of the Parties to the Climate Convention, it does not seem that 'the ethos of democracy can support any norm other than equal per-capita rights to global environmental resources' (Vajpayee, 2002).

Many arguments have been brought forward against the principle of equal per-capita entitlements. None is convincing. The argument that equal per-capita entitlements will create perverse incentives for population growth (see, e.g. Smith et al., 1993, p. 72) overestimates the likely market value of emission rights and underestimates the many serious policy problems that come with population growth. Incentives for recipient Southern nations to push in international negotiations for ever harder global emission caps (which would increase the market value of their excess emission rights) could be reduced through the introduction of double-weighted majority voting as under the Montreal Protocol or the Global Environment Facility. It might well be that the income of developing countries from emissions trading will not be used for climate change mitigation, but for other purposes – but so it is with any marketable resource, such as land, water, or biological resources. Likewise, different developing countries will benefit differently from equal per-capita entitlements – richer countries such as South Africa will benefit significantly less, if at all – yet this also seems justified. Finally, although equal per-capita entitlements of countries ignore the question of unequal per-capita emissions within nations, which differ substantially in many (especially developing) countries, it is doubtful whether governments will ever allow international regimes to address domestic distributive issues. They would be technically difficult to handle in any case.

There remains one key problem with the principle of equal per-capita entitlements: once linked to emissions trading, the principle will create some, potentially significant, financial transfer from North to South. This will be a major incentive for developing countries to join the mitigation regime, but also the key disincentive for the North to accept it. Mixed formulas for a transition period of several decades seem to be the only solution in order to limit North–South transfers to politically acceptable proportions. These could be accompanied by sizeable transaction fees for emissions trading that would be used exclusively to pay for environmental projects (in the South); this would direct some of the transferred resources to global mitigation efforts and still benefit Southern interests. Over time, efficiency gains in the North and economic expansion in the South are likely to result in some degree of converge of per-capita emissions of most nations. In the long run, only poor countries with limited economic activity will remain major net beneficiaries of global emissions trade, which could be compared to them renting out their space and sink capacity to the majority of the more industrialized countries. In sum, climate governance will be successful only if it builds on short-term quantified targets based on a mixed formula that strives towards one long-term principle – equal per-capita entitlements. Europe needs to play a leading role in achieving this goal.

Climate politics

These strategic choices will finally influence European climate politics. Again, the Union is faced with two major negotiating situations: the anti-Kyoto coalition of the USA and Australia, and the group of developing countries.

European climate politics vis-à-vis the USA

European climate politics *vis-à-vis* the USA will develop between two alternatives, conciliation or confrontation. A conciliation strategy would include the preparedness of the European Union to engage with the USA in accepting separate agreements outside the Kyoto framework or to abandon the Kyoto process (after 2012) altogether, and to join the USA in pressuring developing countries to accept a 'meaningful participation' in the climate regime.[23] The aim of a conciliation strategy would be to bring the USA eventually back into the multilateral process, even though probably largely on Washington's terms. This conciliation strategy would entail a number of benefits for the European Union: it would improve transatlantic collaboration and help the climate process through enticing the world's largest greenhouse gas emitter to increase mitigation efforts. On the other hand, bringing the USA back into the multilateral climate framework might require a softening of the overall environmental policy ambition, which could negate gains from increased US efforts. Increased pressure by the EU on the South to engage in 'meaningful participation' might increase the North–South rift in climate policy. Regarding the overall political framework, it might even be in the larger political interest of the EU to resist the temptation of assuaging the USA and instead to assert its own independent role on the global stage by making the Kyoto accord a widely visible success story even without US participation (for more detail, see Biermann and Sohn, 2004).

The latter would require a more confrontational stance and an assertive strategy for making Kyoto work and developing robust incentives for the USA to participate or at least not to obstruct the multilateral effort. One issue that might play an increasing role in the future is the growing gap in the cost of energy in European countries and in the USA, partly due to differences in environmental and energy taxation. Theoretically, such differences in energy taxation could be levelled out through the introduction of border tax adjustments for US products imported into

Europe. I have shown in previous papers (Biermann and Brohm, 2003, 2005 that such border tax adjustments are unlikely to be in violation of world trade law. To some extent, the USA itself has applied similar measures in support of its own taxation of ozone-depleting substances. Border tax adjustments could relate to certain benchmarks of energy content in a given product, and they could be restricted to particular energy-intensive industries or products. The US experience with border tax adjustments on ozone-depleting substances indicates that such schemes can indeed be administered and implemented. Nonetheless, turning energy price differentials between Europe and the USA into a political issue and countering it by levying border taxes on US products (and by rebating European exporters) will result in significant transatlantic conflicts that need to be considered by EU decision-makers.

European climate politics vis-à-vis *developing countries*

European climate politics *vis-à-vis* the developing world, on the other hand, is squarely linked to the overall context of North–South relations, including European external economic policies. In the long term, even massive attempts by industrialized countries to scale down greenhouse gas emissions will not suffice to mitigate climate change if developing countries do not join the effort at some point and reduce their overall emissions. The question, rather, is when, on what terms, and which developing countries will accept costly obligations.

A key condition of the South appears to be that some form of the principle of equal emissions per capita is accepted by relevant actors in the North. In addition, many developing countries link their preparedness to shoulder parts of the burden not only to the details of the commitments to be undertaken and the overall regime structure, but also to possible larger bargains within and outside the climate policy domain. Russian acceptance of the Kyoto Protocol has been interpreted by some observers as being linked to European support for Russia's accession to the World Trade Organization. Engaging developing countries could require similar grand bargains. For one, progress in the climate policy domain could be linked to concessions of the EU in WTO negotiations, for instance when it comes to trade in agricultural products, textiles, or patent rights. For the South, it is key that the North accepts climate change as part of the larger quest for sustainable development, not as a purely environmental problem (see also Najam et al., 2003, p. 228). Second, the climate convention itself links any mitigation efforts of developing countries with the reimbursement of their incremental costs by the North through new and additional resources. Again, the seriousness of the EU's intention to participate in global programmes for the transfer of climate-friendly technology or to support the replenishment of the Global Environment Facility will have a direct influence on the preparedness of the South to assume commitments to at least stabilize greenhouse gas emissions.

Similarly, the main concern of developing countries is not so much the mitigation of climate change – which they expect the North to do, given its higher responsibility and capability – but adaptation (see also Najam et al., 2003, p. 227). Climate science increasingly confirms that it is the South that will suffer most from climate change. Exposure to climate risks in the South may be higher in some regions, and vulnerability in many poorer countries is certainly high. Assistance in adapting to climate change is therefore one key demand that the South directs to the EU, including in the form of an adaptation protocol to the Climate Convention, as proposed by India. Europe needs to respond to this challenge. Preventing large-scale damages through climate change is not only a humanitarian responsibility; it is also politically required. Increasing global interdependence will increase European involvement in the calamities befalling the South brought about by climate change impacts, ranging from floods and storms to drought and land degradation, all of which could result in economic crisis, migration, and political instability. Last, but not least, assisting developing countries in their adaptation efforts will be an important bargaining concession in the overall negotiation of the post-2012 climate regime.

And yet, while mitigation has been intensively studied in both its technical and political dimensions, the emerging need for global adaptation governance is not well understood. This applies both to incremental climate change and to sudden climate change impacts. What will be the best national and international political response to slowly shifting climate zones that will affect sectors as diverse as agriculture, forestry, tourism and human settlement? What political, social and legal criteria will determine the direction of adaptation efforts, since not each and every asset can be equally protected from climate change? How to deal with national, regional or even transcontinental migration due to climate change? The developing countries request the North to assist in finding solutions to these problems, and to shoulder a sizable part of the burden of increasing the adaptive capacity in the South. Europe will have to respond to this challenge.

Conclusion

In climate governance, as well as in other areas of global environmental governance, Europe stands in the middle between the USA and the South. It stands between the adherence to its traditional ally on the other side of the Atlantic – which appears to show decreasing respect for, and interest in, European positions and interests – and the new rising powers in the developing world, notably in Asia, with China, India and the ASEAN. In another paper (Biermann and Sohn, 2004), I have argued that Europe should take this strategic position more seriously and should consciously strive to build up stronger, more stable relationships with the emerging great powers of Asia. In climate governance, Europe is forced to mediate its own interest in the climate issue with a multitude of non-European interests and negotiating positions, but also to forge a coalition of nations that is able to secure a credible, stable, flexible and inclusive governance system for the decades and centuries to come. In this article, I have argued that Europe should take clear principled positions on a number of key issues; in particular the need to have a strong multilateral framework as the sole and core institutional setting for climate policy, and to accept the principle of equal per-capita emissions entitlements as the long-term normative bedrock of global climate governance. Both positions will alienate the USA, and both will make it more difficult for the USA to rejoin the international community on the climate issue.

Yet there might be ways to engage the USA without compromising on the core principles of multilateralism and equal entitlements. One way could be to link the European emissions trading system within the larger Kyoto context with actors and institutions in the USA (Bodansky, 2002a; Danish, 2004). This would require a number of design features, including similarly defined permits, compatible systems for tracking trades, comparably rigorous greenhouse gas emissions monitoring and reporting systems, and comparably stringent compliance systems (Danish, 2004). These requirements are tough but not unfeasible. The recent 'linking directive' of the EU acknowledges the option of linking trading schemes by providing for the possibility of linking the EU trading scheme with that of regional authorities in non-Kyoto countries.[24] In any event, a successful trading scheme in Europe that effectively brings down greenhouse gas emissions in Member States within a reasonable cost range will remain the most powerful tool to convince actors in the USA, and possibly also in the developing world, that climate policy can work. This might be the largest contribution by Europe.

As expressed by a French delegate at the 2003 conference of the parties to the climate convention, the 21st century will be remembered either as the century of climate change suffering and collective irresponsibility or as the century of climate control and the 'maturing of humanity'.[25] Many expect the European Union to lead in this daunting challenge. Future generations might consider the current construction of a climate governance architecture as one of the largest collective institution-building efforts that humankind has ever faced, and on a par with the San Francisco conference that saw the creation of the United Nations Organization. At present, however, it remains to be seen whether the EU will live up to this challenge. The president of the ninth conference of the parties, Miklós Persányi, compared the current institution-building process with the construction of the Milan cathedral, one of the hallmarks of European Gothic art. He did not mention, however, that building this cathedral took the time and energy of five generations of Milanese and no less than 119 years. It is hoped that European construction work on climate governance will require less time.

Acknowledgements

Many thanks to Harro van Asselt, Steffen Bauer, Aarti Gupta, Heike Schröder, participants at a workshop of the European Forum on Integrated Environmental Assessment (EFIEA) in Norwich in October 2004, as well as two reviewers of *Climate Policy* for valuable comments on earlier drafts.

Notes

1 On the debate of the problem-fit of institutions, see Young (2002).
2 This is largely linked to the theoretical strand of sociological institutionalism. See, among many others and with further references, March and Olsen (1998), Finnemore (1996), Hall and Taylor (1996), Barnett and Finnemore (1999), Biermann and Bauer (2005) and Finnemore and Sikkink (1998).
3 On climate policies pursued within Europe, see, e.g., Nilsson and Nilsson (2005).
4 The positions of non-European industrialized countries are reviewed in, e.g., Müller et al. (2003).
5 These smaller Southern country coalitions, all of which harbour views which are not necessarily in conflict with general G-77 positions but can rather be seen as additional, issue-specific special interests, are the Organization of Petroleum Exporting Countries, the Alliance of Small Island States, the Least Developed Countries, and the small Environmental Integrity Group (Mexico, South Korea and Switzerland).
6 Analysed in, for example, Andresen and Agrawala (2001).
7 The Byrd–Hagel Resolution, sponsored by Senator Robert Byrd (D-WV) and Senator Chuck Hagel (R-NE), passed by the Senate with a 95-0 vote, 105th Congress, 1st Session, S. Res. 98.
8 The Clinton administration signed the Kyoto Protocol on 12 November 1998, but did not submit it to the Senate for consideration and ratification. The Bush administration has declared its intention not to ratify.
9 Opening Statement of US Senator Chuck Hagel at the Joint Hearing of the Senate Foreign Relations Committee and the Environment and Public Works Committee, 24 July 2002. Available at http://epw.senate.gov/107th/Hagel_072402.htm.
10 Quoted in Earth Negotiations Bulletin, vol. 12, no. 231, 15 December 2003.
11 Of course, non-ratification does not always imply non-compliance. In particular regarding the USA, international agreements are at times not ratified by the US Senate, but are still largely complied with by the administration. This is not the case, however, for the Kyoto Protocol, which all branches of the current US government reject.
12 This date and comparison was given in the Byrd–Hagel Resolution (based on the assumption that the USA would implement the Kyoto Protocol). See Müller et al. (2003) for a useful discussion of the China–USA comparisons.
13 Bodansky (2002a, p. 6) suggests Colombia, Costa Rica and Mexico, traditionally close allies of the USA and, in the case of Mexico, even a member of NAFTA.

14 See the US-led Carbon Sequestration Leadership Forum, available at http://www.fe.doe.gov/programs/sequestration/cslf/cslf_charter.pdf.
15 For example, CICERO researchers also propose a coalition of most willing nations as 'an interesting supplement to a global UN-based process [...] in particular if the Kyoto Protocol should fail' (Torvanger et al., 2004, p. ii).
16 This includes the agreement of July 2003 of ten north-eastern US states to develop a regional strategy to mitigate carbon dioxide emissions from power plants, including through regional cap-and-trading programmes. The same US region has begun in 2001 to work with Canadian provinces on emission reduction programmes.
17 UN General Assembly Resolution 43/53 of 6 December 1988. See Biermann, (1996, 2002).
18 Informal Working Group of Experts on Financial Mechanisms for the Implementation of the Montreal Protocol (1989, para. 8).
19 Schelling (2002, p. 3), cited in Müller et al. (2003, p. 5-1).
20 One exception of long-term commitments that surpass the tenure of existing policymakers is the granting of long-term loans between governments. Experiences with this system, however, do not bode well for having it implemented in the climate regime.
21 For such a perspective that explicitly argues that 'developing countries should be fully compensated for their emission abatement efforts, but should not receive any further transfers', see C. Böhringer and C. Helm (2004, unpublished).
22 The 1987 Montreal Protocol provides special rights and lesser obligations for all developing countries with a per-capita annual consumption of less than 300 g of controlled substances. This threshold was chosen to be so high that almost no developing country ever crossed it: In 1995, only five developing countries – Kuwait, Lebanon, Slovenia, United Arab Emirates and Cyprus – had to comply with the commitments of industrialized countries, whereas 101 developing countries (including Romania and the successor states of Yugoslavia except Slovenia) fell under the purview of the special regulations.
23 It is too early to tell whether the recent visit of US President George W. Bush to Europe has resulted in a mutual reconciliation on climate policy. The joint action programme of Germany and the USA, agreed upon in February 2005 during Bush's visit to Germany remains vague. While both countries reaffirm their 'history of working together ... to ... mitigate greenhouse gas emissions through such mechanisms as the UN Framework Convention on Climate Change and its Delhi Declaration, the G-8 Action Plan on Science and Technology for Sustainable Development, and the World Summit on Sustainable Development Plan of Implementation', the Kyoto Protocol is not mentioned. Instead, both governments commit to 'broaden and reinforce their activities in three areas of common action to improve energy security and reduce pollution and greenhouse gas emissions, while supporting strong economic growth: First, joint activities to further develop and deploy cleaner, more efficient technologies to support sustainable development; second, cooperation in advancing climate science and developing effective national tools for policy action; third, joint action to raise the efficiency of the energy sector and address air pollution and greenhouse gas emissions in our own countries and around the world' (see German Government, 2005).
24 Environment Daily, 7 April 2004. In order to reach consensus on the proposed directive, the European Parliament was forced to tone down the wording of this provision. Providing explicitly for a possibility to link the EU trading scheme with, for example, the New England/Eastern Canadian trading scheme could undermine the US domestic approach and would probably be an affront to the Bush administration.
25 Quoted in Earth Negotiations Bulletin, vol. 12, no. 231, 15 December 2003.

References

Aldy, J.E., Orszag, P.R., Stiglitz, J.E., 2001. Climate change: an agenda for global collective action. Prepared for the conference on The Timing of Climate Change Policies, Pew Center on Global Climate Change, Washington DC.
Aldy, J., Barrett, S., Stavins, R., 2003. Thirteen plus one: a comparison of global climate policy architectures. Regulatory Policy Program Working Paper RPP-2003-04.
Andresen, S., Agrawala, S., 2001. US climate policy: evolution and future prospects. Energy and Environment 12(2–3), 117–139.
Barnett, M.N., Finnemore, M., 1999. The politics, power and pathologies of international organizations. International Organization 53(4), 699–732.
Barrett, S., 2001. Towards a better climate treaty. World Economics 3(2), 519–541.
Barrett, S., 2002. Towards a Better Climate Treaty. Opinion Pieces, AEI-Brookings Joint Center for Regulatory Studies.
Benedick, R., 2001. Striking a new deal on climate change. Issues in Science and Technology (online), Fall [available at http://www.issues.org/issues/18.1/benedick.html].
Biermann, F., 1996. Common concern of humankind: the emergence of a new concept of international environmental law. Archiv des Völkerrechts 34(4), 426–481.

Biermann, F., 1999. Justice in the greenhouse: perspectives from international law. In: Tóth, F.L. (Ed.), Fair Weather? Equity Concerns in Climate Change. London: Earthscan Publications, pp. 160–172.

Biermann, F., 2002. Common concerns of humankind and national sovereignty. In: Globalism: People, Profits and Progress: Proceedings of the 30th Annual Conference of the Canadian Council on International Law, Ottawa, 18–20 October 2001. Kluwer, Dordrecht, The Netherlands, pp. 158–212.

Biermann, F., Bauer, S., 2005. Managers of global governance: assessing and explaining the influence of international bureaucracies. Global Governance Working Paper. The Global Governance Project, Amsterdam, Potsdam, Berlin, Oldenburg [available at http://www.glogov.org].

Biermann, F., Brohm, R., 2003. Implementing the Kyoto Protocol without the United States: the strategic role of energy tax adjustments at the border. Global Governance Working Paper No. 5. The Global Governance Project, Amsterdam, Potsdam, Berlin, Oldenburg [available at http://www.glogov.org].

Biermann, F., Brohm, R., 2005. Implementing the Kyoto Protocol without the USA: the strategic role of energy tax adjustments at the border. Climate Policy 4, 289–302.

Biermann, F., Sohn, H.-D., 2004. Europe and multipolar global governance: India and East Asia as new partners? Global Governance Working Paper No. 10. The Global Governance Project, Amsterdam, Potsdam, Berlin, Oldenburg [available at http://www.glogov.org].

Bodansky, D., 2002a. U.S. climate policy after Kyoto: elements for success. Policy Brief 15 (April). Carnegie Endowment for International Peace, Washington DC.

Bodansky, D., 2002b. Linking U.S. and International Climate Change Strategies. Pew Center, Washington DC.

Bradford, D.F., 2002. Improving on Kyoto: greenhouse gas control as the purchase of a global public good. Princeton University Working Paper, Draft of 30 April 2002. Princeton, NJ, Princeton University.

Cooper, R., 1998. Toward a real treaty on global warming. Foreign Affairs 77(2), 66–79.

Cooper, R., 2001. The Kyoto Protocol: a flawed concept. Environmental Law Reporter 31, 11484–11492.

Danish, K.W., 2004. Linking a U.S. federal climate program with international and sub-federal climate programs. In: Riggs, J.A. (Ed.), A Climate Policy Framework: Balancing Policy and Politics. Report of an Aspen Institute Climate Change Policy Dialogue, 14–17 November 2003. Pew Center, Washington, DC, pp. 73–82.

Egenhofer, C., Fujiwara, N., 2003. The Emerging Climate Change Regime: The UNFCCC, the Kyoto Protocol Forever, Kyoto Modified … Or Yet Something Else? Final Report of a Study Prepared for Environmental Studies Group Economic and Social Research Institute, Cabinet Office, Government of Japan.

Finnemore, M., 1996. Norms, culture, and world politics: insights from sociology's institutionalism. International Organization 50(2), 325–347.

Finnemore, M., Sikkink, K., 1998. International norm dynamics and political change. International Organization 52(4), 887–917.

German Government, 2005. Press release: US–German joint actions on cleaner and more efficient energy, development and climate change [available at www.bundesregierung.de/Anlage792411/Deutsch-amerikanisches+Aktions-programm+%28engl.+Version%29.doc (last accessed 3 March 2005)].

Gupta, J., 1997. The Climate Change Convention and Developing Countries: From Conflict to Consensus? Kluwer Academic, Dordrecht, The Netherlands.

Hale, T.N., Mauzerall, D.L., 2004. Thinking globally and acting locally: can the Johannesburg partnerships coordinate action on sustainable development? Journal of Environment and Development 13(3), 220–239.

Hall, P.A., Taylor, R.C.R., 1996. Political science and the three new institutionalisms. Political Studies 44(5), 936–957.

Informal Working Group of Experts on Financial Mechanisms for the Implementation of the Montreal Protocol on Substances that Deplete the Ozone Layer, 1989. Report of the Meeting in Geneva, 3–7 July 1989. UN Doc. UNEP/OzL.Pro.Mech.1/Inf.1 of 16 August 1989.

Kant, I., 1795/1983. Zum ewigen Frieden: Ein philosophischer Entwurf. In: von Weischedel, W. (Ed.), Immanuel Kant: Werke in 6 Bänden. Wissenschaftliche Buchgesellschaft, Darmstadt, Germany, pp. 191–251.

Kohler, D.F., Haaga, J., Camm, F., 1987. Projections of Consumption of Products Using Chlorofluorocarbons in Developing Countries. Rand Corporation, Santa Monica, CA.

March, J.G., Olsen, J.P., 1998. The institutional dynamics of international political orders. International Organization 52(4), 943–969.

Müller, B., Drexhage, J., Grubb, M., Michaelowa, A., Sharma, A., 2003. Framing Future Commitments: A Pilot Study on the Evolution of the UNFCCC Greenhouse Gas Mitigation Regime. Oxford Institute for Energy Studies EV 32 (June), Oxford.

Najam, A., Huq, S., Sokona, Y., 2003. Climate negotiations beyond Kyoto: developing country concerns and interests. Climate Policy 3, 221–231.

Nilsson, M., Nilsson, L.J., 2005. Towards climate policy integration in the EU: evolving dilemmas and opportunities. Climate Policy 5, 363–376.

Philibert, C., Pershing, J., 2001. Considering the options: climate targets for all countries. Climate Policy 1, 211–227.

Risse, T., Ropp, S.C., Sikking, K. (Eds), 1999. The Power of Human Rights: International Norms, and Domestic Change. Cambridge University Press, New York.

Schelling, T.C., 2002. What makes greenhouse sense? Time to rethink the Kyoto Protocol. Foreign Affairs 8 (May/June) [available at http://www.foreignaffairs.org/20020501facomment8138/thomas-c-schelling/what-makes-greenhouse-sense.html].

Simpson, B., 2002. Participation of developing countries in a climate change convention protocol. Asia Pacific Journal of Environmental Law 7(1), 39–74.

Smith, K.R., Swisher, J., Ahuja, D.R., 1993. Who pays (to solve the problem and how much)? In: Hayes, P., Smith, K. (Eds), The Global Greenhouse Regime: Who Pays? Science, Economics and North–South Politics in the Climate Change Convention. Earthscan Publications/United Nations University Press, London, pp. 70–98.

Stewart, R.B., Wiener, J.B., 2003. Reconstructing Climate Policy. AEI Press, Washington, DC.

Sugiyama, T., 2003. 'Orchestra of treaties' scenario for after 2012. Presentation for the workshop 'Developing Post-Kyoto Architecture', 5 September 2003, Hamburg [available at: http://www.hwwa.de/Projekte/Forsch_Schwerpunkte/FS/Klimapolitik/PDFDokumente/Taishi.pdf].

Tangen, K., Hasselknippe, H., 2003. Linking treaties to expand the market. Presentation for the side-event 'Developing a post-Kyoto policy framework' at the eighteenth, Bonn, June 12, 2003.

Tol, R.J., 2002. Technology Protocols for Climate Change. An Application of FUND. Working Paper FNU-14.

Torvanger, A., Twena, M., Vevatne, J., 2004. Climate Policy Beyond 2012: A Survey of Long-term targets and Future Frameworks. CICERO Report 2004:02. Centre for International Climate and Environmental Research, Oslo.

Vajpayee, A.B., 2002. Speech by Prime Minister Shri Atal Bihari Vajpayee at the Eighth Session of Conference of the Parties to the UN Framework Convention on Climate Change, 30 October 2002 [available at http://meaindia.nic.in/speech/2002/10/30spc02.htm].

van Asselt, H., Biermann, F., Gupta, J., 2004. Interlinkages of global climate governance. In: Kok, M.T.J., de Coninck, H.C. (Eds), Beyond Climate: Options for Broadening Climate Policy. RIVM Report 500019001/2004 and NRP-CC Report 50003/01. RIVM, Bilthoven, The Netherlands, pp. 221–246.

Victor, David, G., 2001. The Collapse of the Kyoto Protocol and the Struggle to Slow Global Warming. Princeton University Press, Princeton, NJ.

Young, O., 2002. The Institutional Dimensions of Environmental Change: Fit, Interplay, and Scale. MIT Press, Cambridge, MA.

Climate Policy 5 (2005) 291–308

The role of technological development and policies in a post-Kyoto climate regime

Cédric Philibert*

International Energy Agency, Energy Efficiency and Environment Division, 9 rue de la Fédération, 75739 Paris Cedex 15, France

Received 17 January 2005; received in revised form 11 March 2005; accepted 11 March 2005

Abstract

This article discusses various aspects of international technology collaboration and transfer before discussing technology in the climate negotiations themselves. While 'technology agreements' alone could not be expected to deliver the fundamental changes in energy production and use necessary in order to achieve the Convention's objective, they could help in goading developing countries into mitigation action. Other options to speed the diffusion of climate-friendly technologies are considered, although it is unclear whether international climate negotiations offer the best forum. Finally, technology issues should not divert attention from the value of establishing a global emissions trading regime through various options, as this might be a powerful carrier for climate-friendly technologies.

Keywords: Technology; technology agreements; diffusion and transfer; post-kyoto

1. Introduction

Mitigating the effects of climate change will require profound alterations in the way we produce and consume energy. Stabilizing CO_2 concentrations at any level will ultimately require the near-extinction of net CO_2 anthropogenic emissions; and the agenda of slowing, halting, then reversing global energy-related CO_2 emissions will play a primary role in determining the level of CO_2 atmospheric concentrations. Energy efficiency improvements, fuel switching towards less-carbon intensive fossil fuels or non-carbon emitting energy sources such as nuclear and renewable, and carbon dioxide capture and storage all seem useful strategies to achieve the lowest possible concentration level for a given willingness to pay to mitigate climate change (Philibert, 2003).

There is, however, an ongoing debate on the role of 'existing' and 'breakthrough' technologies required to achieve these changes while continuing to fuel the world economy. Hoffert et al. (2002) state that only 'breakthrough' technologies can fuel the world economy with enough energy,

---

* Corresponding author. fax: +33-1-40-57-67-39
E-mail address: cedric.philibert@iea.org

while at the same time reducing GHG emissions; on the other hand, Pacala and Socolow (2004) argue that existing, already proven technologies can do the job. Although significant emission reductions below business-as-usual trends could be achieved in the short term with a wider dissemination of existing technologies, this wider dissemination itself will both require and facilitate incremental improvements that would reduce the costs and increase the performances of these existing technologies – while competing fossil fuel technologies will respond to the competition from new technologies. In any case, deeper changes will be necessary in the longer run in order to facilitate more substantial emission reductions.

Policies specifically designed to promote technical change, or 'technology push', could play a critical role in making available and affordable the new energy technologies that will be required, given the depth of the emission cuts necessary to fulfil the Convention's objectives. Although other instruments, notably market instruments, would also drive some technical change, the usual short-sightedness of market actors is unlikely to provide for the broad changes required.

However, it is equally unlikely that an approach limited to 'technology push' would be sufficient to achieve the Convention's objective, for two reasons (roughly similar but distinct with respect to timing). First, there is a large potential for cuts that could be achieved in the short run with existing technologies; and second, development of future new technologies requires a market pull as much as a technology push, particularly as market pull may speed technology improvements, thanks to learning-by-doing processes (Philibert, 2005).

This article discusses two international dimensions of governments' role in promoting technical change. The first relates to the role of international collaboration on technology innovation and development; the second relates to technology dissemination and transfer. A third section aims at providing insights on the role that technology might play in future negotiations on climate change.

Combining 'push' and 'pull' policies would help bring new climate-friendly technologies into markets and benefit from learning-by-doing processes. Technology pull is also important, as short-term emission reductions determine the level at which atmospheric CO_2 concentrations may be stabilized. Governments can use various tools to promote technical change, as illustrated by the case of the EU (see Box 1).

Box 1: The EU situation

The EU action to mitigate climate change is multifaceted. Its most visible element is its emissions trading programme, although this only covers a part of total industrial emissions – including those from the power sector. Many other coordinated European policies also contribute to this objective, although they may have been adopted in wider contexts and pursue other objectives as well, such as energy security or improvements in the environment at local levels.

Some of these programmes are more than 10 years old but have only progressively been translated into effective measures. A good example in this category might be the 22 September 1992 Directive on labelling of the energy consumption of household appliances – followed by implementing measures from 1994 onwards, starting with refrigerators and freezers and progressively covering all kinds of appliances. Older programmes have also been repackaged and renamed; for example, the 'Save', 'Altener', 'Steer' and 'Coopener' programmes on, respectively, energy efficiency, renewable energy, fuel diversification in transport, and promotion of renewable energy and energy efficiency in developing countries, are now part of an integrated multi-year programme called 'Intelligent Energy for Europe'.

Amongst the most significant recent initiatives are the following:

- the 6 November 2001 Regulation on a Community-wide energy-efficient labelling programme for office equipment ('Energy Star') that followed the conclusion of an agreement with the US government on coordination of labelling programmes;
- the 27 September 2001 Directive on the promotion of electricity from renewable energy sources in the internal electricity market that sets a global EU target for renewable energy at 22.1% of electricity or 12% of gross domestic energy consumption in 2010;
- the 16 December 2002 Directive on the energy performance of buildings;
- the 8 May 2003 Directive on the promotion of the use of biofuels, which sets a minimum of 5.75% of biofuels to replace diesel or petrol for transport purposes by December 2010;
- the February 2004 decision to establish a mechanism for monitoring Community greenhouse gas emissions as an implementation of the Kyoto Protocol's requirement, in advance of its entering into force at international level.

Also worth mentioning is the agreement reached in July 1998 between the Commission and the European Carmaker Association to reduce specific CO_2 emissions of new cars by 25% in ten years, with an objective of an average 140 gCO_2/km in 2008 – followed one year later by an agreement by the Japanese and Korean carmakers' associations to do the same in respect of cars sold in the European Union. In 2004, the Commission reported that European and Japanese carmakers were on track in meeting their commitments.

Obviously the variety of these measures and actions can be partly attributed to the limited coverage of the EU emissions trading regime. One must note, however, that some do interfere with emissions trading, notably the directive on renewable and, to some extent, the demand-side management measures affecting electricity consumption. Among the possible ways for electric utilities to comply with their CO_2 targets should be included DSM programmes and factoring in more electricity for renewables.

Behind these interferences different decisions have been made in the pursuit of various objectives by different parts of the European institutions; the coexistence of emissions trading and the renewable electricity directive can also be seen as a deliberate attempt to correct the myopia of market actors. While perfect static cost-effectiveness can be achieved through plain economic instruments, the search of a more dynamic cost-effectiveness in solving a long-term problem may imply accepting the need to pay more for some emission reductions than for others, with a view of taking advantage of the learning-by-doing processes to reduce the costs of young technologies – in this particular case of wind power and other renewables. This could have been one rational explanation underlying the EU's quest for 'supplementarity' in negotiating the implementation of the Kyoto Protocol's flexibility mechanisms.

While the EU is certainly at the forefront of implementing the Kyoto Protocol – despite the current weakness of the national allocation plans of its emissions trading regime – it might, however, have been less successful in establishing technology partnerships with other countries than the USA.

2. International technology collaboration

International cooperation on R&D allows each participant to benefit from others' efforts. It thus magnifies and accelerates results and helps disseminate them. The international exchange of lessons learned and best practices in energy technology policies, from research to deployment, can contribute to improve the economic efficiency of increasing the use of clean and efficient technologies. Specifically, international collaboration on energy technology development reduces the costs of R&D by enabling the sharing of results and by avoiding the duplication of efforts, and through the resulting increased rate of technological progress. International science and technology development, exchange and diffusion through public–private partnership for technology absorption, capacity building, and innovative project-financing mechanisms are desirable and mutually beneficial to both developed and developing countries.

Cooperation between countries should not preclude competition between companies, and may drive governments to increase their efforts, especially in supporting basic research and development. The reason stems directly from the economic theory of public goods. Rational economic agents acting in isolation are likely to limit their R&D efforts so as to equalize the marginal abatement costs they incur with the marginal benefits they gain from their own efforts. Economic agents cooperating in R&D efforts are likely to raise their level of effort so as to equalize their marginal costs with the benefits that each of them receives from the actions of all. Simply replace 'economic agents' with 'countries' – and the usual justification for public R&D efforts or support (public funding of private R&D) seems to apply equally well to justifying international R&D cooperation.

Increased technology cooperation between countries could help engage more countries into action to mitigate greenhouse gas emissions. Countries reluctant to adopt binding quantitative commitments on their emissions that may be felt as unduly 'restrictive' in terms of economic development might find in technology cooperation a more 'positive' way of introducing – at least in part – the same kind of changing patterns. We discuss in the next section whether 'issue linkage' could help to encourage such countries to adopt targets.

International technology collaboration may have a number of benefits. However, there might be several drawbacks as well. Here are some possibilities:

- The search for wide agreement may require time- and resource-consuming efforts, and may thus slow, rather than speed, the innovation and diffusion process;
- Some players might deliberately slow processes in some technologies in order to protect vested interests in competitive technologies;
- The difficulty of protecting intellectual property rights in close cooperative work may offset an incentive for some players;
- Premature technology selection could impede necessary competition between various technology options.

2.1. The IEA experience

Since its creation in 1974, the International Energy Agency (IEA) has provided a structure for international cooperation in energy technology research and development (R&D) and deployment. The IEA does not do R&D nor finance it. However, it brings together experts in specific technologies who wish to address common challenges jointly and share the fruit of their efforts. Within its structure, there are currently more than 40 active programmes, known as the IEA Implementing Agreements (IA). Almost three decades of experience have shown that these Agreements are contributing significantly to the achievement of faster technological progress and innovation at lower costs. Such international cooperation helps to eliminate technological risks and duplication of efforts, while facilitating processes such as harmonization of standards. Special provisions are applied to protect intellectual property rights.

Working rules are based on the *IEA Framework for Implementing Agreements*, which was adopted in 2003, as a result of the willingness of Ministers from IEA member countries to expand the energy dialogue with key OECD non-member countries and to encourage industry to participate in energy research, development and demonstration (Philibert, 2004a).

Thirty-seven countries participate in 42 Implementing Agreements. Six IAs concern fossil fuel energy, including clean coal and carbon dioxide capture and storage. Nine IAs relate to nuclear fusion power. Eight agreements promote R&D on renewable energy sources, as well as hydrogen production and use. Fourteen IAs are devoted to end-use efficiency improvements in transportation, industry and building. The former 'Climate Technology Initiative' is now one of these Implementing Agreements. An extensive description of the work conducted under Implementing Agreements can be found at: http://www.iea.org/Textbase/techno/index.asp.

2.2. The many bilateral agreements

Bilateral cooperation programmes are numerous – especially (but not exclusively) regarding technology transfer – and many examples can be found in national communications under the UNFCCC. UN agencies, such as UNDP, UNEP, UNIDO and others, as well as other multilateral agencies, such as the World Bank or the OECD, notably through its Development Assistance Committee, are also deeply involved in the transfer of technologies as is, though in a different manner, the World Trade Organization.

In the midst of bilateral and multilateral programmes there are also some relevant plurilateral arrangements; these do not gather almost all countries, as multilateral arrangements usually do (or are intended to do), but a more limited number, although always more than two. G8 is probably the most well-established 'plurilateral' institution, and has considered climate change in various meetings and from various angles. For example, in 1999 the G8 Cologne Summit asked the Export Credit Agencies to harmonize their environmental policies and help developing countries address the challenge of climate change – as did the OECD Council at ministerial level in the same year. At the Evian Summit in 2003, the G8 went much farther and adopted an action plan on '*Science and Technology for Sustainable Development*'. This document contains an important section devoted to energy technology.

Many examples of successful international technology collaboration can be found in other areas than climate change; one such example is that of the Consultative Group on International Agriculture Research (CGIAR), described by Gagnon-Lebrun (2004).

2.3. The recent US initiatives

The US Climate Technology Partnership (CTP) was launched in 2001 to build on the Technology Cooperation Agreement Pilot Project (TCAPP) that began in 1997. This programme supports the implementation of climate-friendly energy efficiency and renewable energy technologies in developing and transition countries. The USA is also part of two important international partnerships: Generation IV International Forum (July 2001), a multilateral partnership fostering international cooperation in research and development for the next generation of safer, more affordable, and more proliferation-resistant nuclear energy systems; and the Renewable Energy and Energy Efficiency Partnership (Johannesburg, August 2002), which seeks to accelerate and expand the global market for renewable energy and energy-efficiency technologies.

Since then the US administration has launched several 'high-level' initiatives and partnerships aiming at collaborating with selected industrialized and developing countries. These are as follows:

- *The International Partnerships for a Hydrogen Economy* (April 2003) to advance the global transition to the hydrogen economy, with the goal of making fuel-cell vehicles commercially available by 2020. The Partnership will work to advance research, development, and deployment of hydrogen and fuel cell technologies; and develop common codes and standards for hydrogen use.
- *The Carbon Sequestration Leadership Forum* (June 2003) to advance technologies for pollution-free and greenhouse gas-free coal-fired power plants that can also produce hydrogen for transportation and electricity generation.
- *The Methane-to-Markets Partnership* (July 2004) will work closely with the private sector in targeting methane currently wasted from leaky oil and gas systems, from underground coal mines, and from landfills.

What these various initiatives are likely to add to existing international collaborative frameworks on the very same topics is probably the awareness-raising and implication of significantly higher-level policymakers in the member countries. This will certainly help speed development and implementation of carbon dioxide capture and storage activities. It remains to be seen whether this will accelerate to any great extent the development of a hydrogen economy, which seems to rest primarily on a large number of scientific and technical breakthroughs, and it also entails the risk of diverting policymakers' interest in less radical, but more practical, options for the coming decades.

3. Technology diffusion and transfer

3.1. The globalization context

Trade liberalization and the globalization of finance have significant implications for climate change policies. As Charnovitz (2003, p.141) noted

> On the one hand, lowering trade barriers and opening markets boosts economic growth, which tends to increase GHG emissions. On the other hand, bigger markets spur technological innovation and diffusion, which can reduce the GHG intensity of economic growth. Moreover, as trade promotes higher national incomes, some countries will find themselves better able to afford emission abatement efforts.

Therefore, independent of any technology cooperation and transfer policy, technologies diffuse from country to country through a wide number of channels, including trade, foreign direct investment and patent licensing, but also emigration, travel and visits, exchanges of students, international scientific and technology journals and conferences, the Internet and others.

There are three basic ways for a firm to exploit its technologies abroad, and consequently, three different ways for countries to acquire that technology:

- *Through trade*: international technology transfer through trade occurs when a country imports higher-quality (than it can produce itself) intermediary goods to use in its own production processes.
- *Through licences*: a firm may licence its technology to an agent abroad who uses it to upgrade its own production.
- *Through investment*: a firm can set up a foreign establishment to exploit the technology itself. Foreign direct investment (FDI) is the most important means of transferring technology to developing countries. Technology transfer through FDI generates benefits that are unavailable when using other modes of transfer. For example, an investment is not only composed of technology, but also includes the entire 'package', such as management experience and entrepreneurial abilities, which can be transferred by training programmes and learning-by-doing. (Tebar Less, 2003).

As official development assistance stagnated, private flows to developing countries have increased roughly five-fold over the past decade – and decreased in both 2001 and 2002 after the 2000 peak (Heller and Shukla, 2003). Foreign direct investment (FDI) took the lion's share over other forms of external financing such as bank lending, but flowed disproportionately to a small number of developing countries. In 2002, developed countries received about three-quarters of total FDI inflows (US$460 billion), while developing countries received one-quarter (US$162 billion) – although China became the first recipient country.

3.2. The export credit agencies (ECAs)

Most energy-related foreign direct investments in developing countries – a substantial share of the total – have some support and guarantees from one or several investors' country export credit agencies (ECAs). These have come under increased scrutiny by analysts, environmental or development NGOs, and multilateral organizations.

As an example, Maurer and Bhandari (2000, p.1) considered that the

> failure to place ECAs within a wider development and environmental context is generating a policy perversity. [...] ECA financing to developing countries favours exports and investments that disproportionately benefit energy- and carbon-intensive industries.

In fact, 40% of foreign investments in developing countries went into fossil-fuelled power generation and oil and gas development from 1994 to 1999, and up to 60% to 'energy-intensive' projects, including manufacturing and transport infrastructures. While Maurer and Bhandari (2000, p.1) fairly recognize that these investments are 'likely upgrading infrastructures, introducing more energy-efficient technologies, and permitting fuel switching from coal to less-intensive natural gas', they believe that 'from a climate perspective, ECAs appear to be doing more harm than good'.

Environmental NGOs, such as Friends of the Earth, went one step further, asking international financing institutions (ECAs and multilateral development banks) to stop financing investments in the fossil fuel sector, including oil and gas pipelines. Some NGOs from developing countries, however, have adamantly criticized this:

By forcing developing countries to stop using fossil fuel technologies without providing a framework for the World to move towards a renewable energy future, Northern groups (i.e., NGOs) are denying these countries the right to development (Sharma, 2000).

It would certainly be counterproductive to 'stop fossil fuel investments in developing countries' from both climate and – of course – development perspectives. On the other hand, the great leverage power of international finance institutions could be better targeted to ensure better coherence of ECAs activities with global sustainable development goals, as suggested by Sussman (2003). She identifies several policy options that might be used to further influence the decisions of ECAs, including (1) a pool of concessionary financing funded by donor contributions; (2) financial set-asides; (3) special lending provisions; (4) climate-friendly portfolio standards with credits and charges; and (5) increased transparency in financial and emissions reporting by ECAs. For most of these options, concessionary financing appears to be key to turning technologies that are not commercial today into viable projects that are consistent with ECA financing rules.

Under the UNEP Finance Initiative, nearly 300 financial institutions in the world – bankers, insurance and assets managers – have undertaken to use their leverage power in support of sustainable development.

3.3. Emission leakage vs. technology spillover

What will be the effects on global emissions of the Kyoto Protocol, now likely to enter into force soon? An assessment limited to emission reductions in industrialized countries having ratified the Protocol suggests that dispositions of the Marrakech Accords and, moreover, the withdrawal of the USA will allow the emissions of Annex I countries together to increase by 9% above 1990 levels (IEA, 2002). Globally, this would correspond to limited reductions below business-as-usual trends, in the order of a few percentage points. However, a fuller assessment must take into account the possible effects on emissions in countries not covered by Kyoto targets, resulting in the opposite effects of emission leakage and technology spillover (not to mention the fact that Kyoto is only a first step).

Assessments of leakage rates range from 5% to 20% in the case of the Kyoto targets (Hourcade and Shukla, 2001) – but could go much higher with larger emission reductions in the future. Against this, Grubb et al. (2003, p.20) set various sources of positive spillover, especially 'the international diffusion of more efficient and lower carbon technologies that are developed in response to emission controls in the industrialized world'. They believe that these positive effects do more than offset the leakage: the Kyoto Protocol with spillover leads to convergence of emission intensities by 2100 and keeps atmospheric concentrations below 560 ppm – while without spillover effects it would not prevent atmospheric concentrations from reaching 730 ppm. Therefore, regulation of emissions by industrialized countries would also reduce emissions – in comparison to business-as-usual trends – in non-regulated areas. This result may seem very optimistic. However, although the aggregate degree of spillover is uncertain, the available evidence suggests that it will be important and environmentally beneficial in aggregate. In the course of industrialization, newcomers tend to see their energy intensity peak at lower levels than countries which have been industrialized for longer (Martin, 1988).

Spillover will help to spread the global effectiveness of the Kyoto first period and subsequent commitments, and deserves much more thorough scrutiny. Positive spillover effects could be greatly enhanced through technology transfer policies and/or mechanisms such as the clean development mechanism or emission trading with developing countries.

3.4. Intellectual property rights

The protection of intellectual property rights (IPR) is often mentioned as an important 'barrier' to technology innovation and diffusion by both industry people and developing country representatives. However, the problem seems to be very different depending from what side it is seen. Developing country representatives often argue that IPR protection tends to slow the diffusion of new climate-friendly technologies in developing countries by making them costlier. Industry people make a quite different case: they often say that they are not inclined to transfer their proprietary technologies to developing country counterparts because they fear that their IPR will not be duly protected. They fear that their technology would be copied without due retribution; moreover, they fear that developing country companies could thus get an unfair competitive advantage on both developing and developed country markets.

The trouble with reducing IPR protection – for example through shorter and narrower patents – is that this would perhaps speed diffusion but at the same time slow innovation. Any IPR protection regime is a compromise between innovation and diffusion and must strike a balance between them both. A stronger protection of rights, provided, for example, by wider and longer patents, would give a greater incentive to innovate but further slow diffusion of innovation. There is a continuous debate on the adequacy of the current international regime. The current compromise may or may not be the optimum one, but clearly climate change mitigation will require both innovation and diffusion of innovation and thus may not be an argument to modify the current regime in one sense or another. Moreover limiting the discussion to this compromise may miss the point of its effective implementation in developing countries. This seems to be a matter of law implementation and governance.

The provision of the Trade-Related Aspects of Intellectual Property Rights (TRIPs) Agreement stipulate that 20-year patent protection should be available for all inventions, whether products or process, in almost all technology fields. This may, or may not, be the best trade-off between the need to reward innovation and the need to speed diffusion. Though economists continue debating the matter, no clear case appears for revising these dispositions.

The TRIPs Agreement, however, explicitly directs developed country members to provide incentives to enterprises and institutions to promote and encourage technology transfer to least-developed countries (Article 66.2). The WTO Fourth Ministerial Conference held in November 2001 in Doha reaffirmed that these provisions are mandatory and requested the TRIPs Council to monitor its implementation.

With respect to the costs of IPR protection in developing countries, it may be useful to recall that the Agenda 21, in its article 34.18, suggests, *inter alia*,

> ... the purchase of patents and licenses on commercial terms for their transfer to developing countries on non-commercial terms as part of development cooperation for sustainable development, taking into account the need to protect intellectual property rights.

Finally, protection of intellectual property rights may not only be necessary to alleviate the concerns of industry people from industrialized countries about selling their technologies. Perhaps even more importantly, it would be needed to alleviate concerns of industry people from developing countries about acquiring technologies. The latter, not the former, would be the primary losers if their competitors were to copy the technologies without paying the cost, as has been shown to be the case with clean coal technologies in china (Philibert and Podkanski, 2005).

4. Technology in the future negotiations

This section aims at investigating what useful role 'technology' could play in the climate negotiations. It first suggests that agreements focusing on technology would be unlikely to be sufficient – or politically credible enough – to deliver the changes required to achieve the objective of the Convention. It then examines the case for linking agreements on technology with adoption of quantitative targets. Harmonization of standards could play a role, but it remains unclear whether the climate negotiations would be the best place. Agreements to share the necessary learning investments in new technologies could be useful as well. Moving from the Clean Development Mechanism through plain emissions trading, thanks to more flexible commitment options, may remain an essential element of a comprehensive strategy to 'pull' as well as to 'push' climate-friendly technologies into the markets.

4.1. Technology agreements: can they deliver?

A number of recent proposals have been made for technology-based international agreements as successors to the Kyoto Protocol. Scott Barrett (2003) suggests the negotiation of a new agreement focusing on R&D funding. While such an agreement might complement the current Kyoto Protocol, Barrett maintains that over time it could fully replace it. Under his proposal, base-level contributions would be determined on the basis of both ability and willingness to pay, and could be set according to the UN scale of assessments. To provide incentives for participation, each country's contribution to the collaborative effort would be contingent on the total level of participation. The research emphasis would be on electric power and transportation. This would be a 'push' programme for R&D – a dimension absent from the Kyoto approach.

However, Barrett also proposes a complementary 'pull' incentive to encourage compliance and participation. He suggests that the most attractive approach would be to agree on common standards for technologies identified by the collective R&D effort, and established in complementary protocols. As examples, energy efficiency standards could be established for automobiles, requiring the use of new hybrid engines or fuel cells, or standards for fossil-fuel-fired power plants might require capture and storage. Barrett's proposal also includes a multilateral fund to help spread the new technologies to developing countries, a short-run system of pledge-and-review, and a further protocol for adaptation assistance.

A standards-based approach was also advocated by Edmonds (1999, 2002) and Edmonds and Wise (1999). Under their hypothetical protocol, any new fossil fuel electric power plant and any new synthetic fuels plant installed in industrialized countries after 2020 would be required to capture and dispose of any carbon dioxide from its exhaust stream or conversion processes. Developing countries would undertake the same obligations when their per-capita income equals the average for industrialized countries in 2020 in purchasing power parity terms.

The most problematic aspect of such a strategy might be one of credibility – a problem inherent in approaches based on still-to-be-developed technologies. The difficulty in setting standards is that regulators do not know the exact amount of improvement that is feasible; standards thus run the risk of being either too lax or ultimately unachievable. To make things worse, companies often anticipate that political authorities will waive the target if technological improvements are insufficient, particularly if the consequences would be costly and politically difficult. The dynamics of incentives for innovation is unclear in such a case. While there remains a strong incentive for innovators, companies subject (even indirectly) to the regulation might prefer not to develop the appropriate technologies and wait for the authorities to waive the target. And in some cases – as arguably in the automobile industry – only companies potentially subject to regulation have the financial and technical resources to fully develop radically new technologies.

No less important is the cost issue. Edmonds and Wise (1999) themselves recognize that the cost of achieving a given concentration level with such a protocol would be 30% higher than the economically efficient cases of taxes or tradable permits. This estimate may even be too low, as the structure of the agreement would not encourage some of the most cost-effective energy efficiency improvements. In addition, the politics of some technology proposals may make them difficult to implement – particularly if they tend to disadvantage specific (and politically powerful) – segments of the economy. Thus, for example, a technology proposal that calls for phasing-out coal may meet the same experience as that faced in England or Germany, where closing down even money-losing coal mining operations is a process that takes decades.

Barrett (2003, p.395) also recognizes that his approach would not be cost-effective and thus would only be a second best. But, he argues, the setting of standards

> ... often creates a tipping effect. If enough countries adopt a standard, it may become irresistible for others to follow, whether because of network effects, cost considerations (as determined by scale economies), or lock-in.

Well, it may ... or may not. Let us suppose that some industrialized countries adopt a standard that would, for example, force energy-intensive industries, the power sector, and refineries to give up fossil fuels or capture and store the carbon dioxide. Its not easy to figure out why this would obligate or incite the rest of the world to follow, especially if this entails huge costs.

Would new multilateral funds make the difference? Maybe – but it is not obvious that new funds leveraging only scarce public money would do more than mechanisms, such as emissions trading, leveraging potentially both public and private money. Also, if some of these technologies become fully cost-effective, thanks to economies of scale and learning curves, then they might be disseminated by their own virtues. The technology spillover effects might be similar to the Kyoto case. Finally, the Intergovernmental Panel on Climate Change (IPCC) made it clear that energy efficiency improvements at the end-user level, likely to provide the bulk of short-term affordable emission reductions, require 'hundreds of technologies' (Moomaw and Moreira, 2001). Should one then negotiate hundreds of protocols?

4.2. Linking technology cooperation and targets: will it work?

From a theoretical standpoint, extensive cooperation on global environmental issues is difficult to achieve because the public nature of the global environment creates strong incentives to free-ride. The literature often suggests 'issue linkage' as a way to overcome such difficulties. Buchner et al. (2003) have investigated whether a linkage between cooperation in the field of low-carbon technology R&D and cooperation in emissions control could provide an incentive for the USA to 'come back' to the Kyoto Protocol. The incentive would thus be, for the USA, to benefit from technological spillover arising from R&D efforts undertaken by others – namely the EU, Japan and the Former Soviet Union in this study. Their conclusion is that such issue linkage would not be an effective strategy to induce the USA to 'come back' to emissions control, because it would be based on an implicit non-credible threat. By refusing to cooperate with the USA on technology, the EU, Japan and Russia would simply increase their losses: 'they would thus 'prefer to cooperate with the US on technological innovation and diffusion even when the USA free-ride on climate cooperation' (Buchner et al., 2003, p.23). One key factor explaining this result may be the high R&D expenditures level in the USA, making technology cooperation profitable to others. However, this analysis also suggests that all Annex I countries would benefit from technology cooperation.

For other countries, notably most developing countries, an 'issue linkage' between technology cooperation and emission control may prove effective and induce more countries into taking more action. Taking an example from the Montreal Protocol, Benedick (2001) writes that 'technology provides an irresistible incentive for developing countries to accept commitments'. Of course, some experts in developing countries may argue that such a linkage would be unfair, as technology transfer is a commitment of developed countries under the Convention. However, Article 4 commits all Parties to 'formulate, implement, publish and regularly update national and, where appropriate, regional programmes containing measures to mitigate climate change'. Moreover, such a linkage is already embedded in the Convention Article 4.7.

Technology cooperation may ease some of the barriers to strengthening emissions mitigation cooperation even if no direct linkage is made between the two. This could be achieved by (a) promoting a deeper understanding of each other's difficulties; (b) helping build confidence among countries; (c) increasing the depth of relationship between government, NGO and business people in and between the various countries; and (d) remaining engaged on common mitigation action at a time when countries have difficulties agreeing on any global scheme to address greenhouse gas emissions.

4.3. Technology standards: the benefits and risks of harmonization

However, there are a number of areas where standards could prove efficient, cost-effective – and international collaboration could help promote them. One such example is the least-lifecycle-cost strategy for energy efficiency improvements in the area of domestic appliances that has been suggested by the IEA (2003). Based on a least-lifecycle-cost assessment for most domestic appliances, effective policies in OECD countries could result in 20 years' emission reductions of about 322 million t CO_2/year by 2010, 470 Mt CO_2/year by 2020 and 572 Mt CO_2/year by 2030 – or roughly 30% of OECD member countries' targets under the Kyoto Protocol. This would be achieved at no cost for the society and consumers – or, more precisely, a cost for government but much larger benefits for consumers. Figures would be even higher if the 'least lifecycle cost' were to include a price for avoided carbon emissions (and/or other positive externalities).

Effective policies would use a variety of means, from awareness-raising campaigns to procurement programmes, so as to give appropriate incentives at all relevant levels, from manufacturers to retailers and other 'market intermediaries' to end-use consumers. Most likely, these policies would make a significant use of labels or minimum efficiency standards or both – which have already been proved effective.

With increasing globalization of appliance and technology markets, however, international cooperation on appliance policy is becoming an essential element of product markets. It can generate greater transparency and comparability in appliance standards, test procedures and labelling, which would bring benefits for producers, consumers and governments alike. It would reduce costs for product testing and design, enhance prospects for trade and technology transfer, and assist governments and utilities in efforts to design, implement and monitor efficiency programmes.

It is not clear, however, to what extent international cooperation should aim at harmonization. The IEA (2003) sees harmonization of test protocols as almost always positive, while harmonization of labels and standards offer a more variegated picture. Differences in climate, electricity prices, consumers' attitudes and other factors may make harmonization very difficult, but also meaningless in some cases. Harmonization of labels and standards makes most sense for products whose characteristics and usage patterns do not vary greatly from country to country and where the level of efficiency economically justifiable is somewhat insensitive to energy prices.

Neither harmonization nor even collaboration need always be global. On the contrary, the search for exhaustive geographical coverage may prove counterproductive in delaying real work and outputs. The evaluation of the trade-off between rapidity and exhaustiveness will probably have to take into account some spillover effects – as happens when a sufficiently large number of countries adopt standards that soon become world class standards. This has been the case with many safety standards in various domains. Finally, while governments could be willing to strengthen cooperation in this field, bringing this topic into the climate negotiations is not necessarily the best way forward.

4.4. Sharing the necessary learning investments

New technologies in general witness learning effects on costs: as their markets expand, costs go down (IEA, 2000). Consequently, it is hoped that expanding the markets for technologies already not too far from competitiveness could expand their niche markets towards full competitiveness – or at least competitiveness in markets where externalities are properly priced – within a decade or two. Achieving this is likely to require, alongside sustained R, D&D efforts in public–private partnerships, a somewhat subsidized or 'artificial' expansion of markets. The total amount of these 'subsidies' (which could take the form of feed-in tariffs in the electricity sector) would represent 'learning investments'.

The wind power industry has benefited from such efforts in the USA in the 1980s, followed by Denmark, then Germany, Spain, the UK, and progressively more countries, including India. Concerted efforts might be more productive – and the costs of future similar 'learning investments' could be shared by a broader set of countries willing to cooperate. Other renewable technologies, from PV to concentrating solar to biomass power to offshore wind, may benefit from such international cooperation – which would go beyond the R&D shared efforts in the IEA's implementing agreements (see, e.g., Philibert, 2004b, for a case study on concentrating solar power technologies).

This is what the GEF (Global Environment Facility) is trying to do with its Operating Programme No. 7, but the irony is that it does so with developed country money only in developing country fields. It is hard to see why such efforts are not better connected to similar efforts being undertaken in various developed countries through renewable energy portfolio policies.

In the lead-up to the World Summit on Sustainable Development, a proposal for all countries to commit themselves to reach some agreed percentage of renewable energy sources in their primary energy supply was partly intended to accelerate learning-by-doing from accelerated deployment for these necessary technologies. Although the EU and some developing countries have supported this proposal, other industrialized countries and most developing countries opposed it.

This proposal, discussed at Johannesburg, was expressed in percentage terms. This suggests that either reducing energy consumption or increasing renewable energy supply could have been used to achieve these commitments. It is possible that a more focused approach, which would be expressed in absolute, not relative terms, offers better prospects for reaching a consensus. Metrics might be capacity, energy, or even physical units such as square metres (say, for PV, or solar heating). If the purpose of such approach is to foster renewable energy deployment, not allowing flexibility with energy efficiency improvements would better serve it. It may also be that 'plurilateral' agreements on fostering renewable energy sources between like-minded countries offer better prospects for a prompt start – with the hope they be broadened progressively.

This may still leave open the question of integrating efforts undertaken in developing countries with GEF and other developed-country financing and those undertaken in developed countries themselves. While this may be a topic for negotiators in the UNFCCC, initiatives may also be part of the implementation of the G8 Evian Action Plan for Sustainable Development.

4.5. Moving into plain emissions trading

The various funds created in the course of the climate negotiations all rest on public money and are unlikely to be sufficient to drive the fundamental changes needed in developing countries. The Clean Development Mechanism might also be intrinsically limited. Moving into plain emissions trading might be the best hope for far-reaching changes but would require new options for future targets.

4.5.1. The GEF and the SCCF

Since its inception, the Global Environment Facility (GEF), acting as the financial mechanism of the Convention, has given US$1 billion for climate change projects and leveraged more than US$5 billion in co-financing. Usually, GEF funds are linked to other loans from multilateral institutions (e.g. the IBRD or the ADB); they are also linked to projects financed with national or bilateral funds. More than half has been devoted to renewable energy projects and more than a quarter to energy efficiency projects in 47 developing and transitional countries. The 2002 repartition gave greater emphasis to enabling activities (12.3%) and sustainable transportation (4.4%), while the energy efficiency projects rose almost to the level of renewable energy projects, with 40.3% and 42.3% of funding, respectively.

In 2002, just before the World Summit on Sustainable Development in Johannesburg, donor nations agreed to replenish GEF's trust fund by US$3 billion – the largest amount ever. The funds will be spent between 2002 and 2006 on the four initial GEF topics (biodiversity, climate change, international waters, and the ozone layer) plus two new ones (land degradation and persistent organic pollutants).

The GEF Operating Programme No. 7 is of particular interest, built around the notion of learning-by-doing. One of its objectives is the reduction in costs of low greenhouse gas emitting technologies by increasing their market shares.

The Marrakech Accords created a Special Climate Change Fund (SCCF) under the Convention to provide additional assistance for adaptation, technology transfer, energy, transport, industry, agriculture, forestry, and waste management, and broad-based economic diversification. A Least Developed Countries Fund was also established under the Convention, while an Adaptation Fund was established under the Kyoto Protocol. Previously in Bonn, at COP-6, a number of countries (the EU, Canada, Norway, New Zealand, Switzerland and Iceland) made a political statement that they would provide a minimum of US$410 million per year for climate change activities – including their climate contribution to the GEF. Recently, the Parties to the Convention agreed that the Special Climate Fund should serve as a catalyst to leverage additional resources from bilateral and multilateral sources. There was also agreement that top priority should be given to funding of adaptation activities to address the adverse impacts of climate change from the resources of the SCCF; and that technology transfer and its associated capacity building was also important.

However, these new funds were created in a context of increasing scarcity of public spending in most OECD countries. Many observers believe that while it will help to finance capacity building at national levels, it will never be large enough to finance the costs associated with the profound changes in the energy sector required to promote development while reducing global emissions. This scepticism is also fuelled by the decline in Official Development Assistance.

4.5.2. The CDM and why it cannot do much

To some extent, however, the Clean Development Mechanism (CDM) instituted by the Kyoto Protocol may substitute for quantified objectives by developing countries and give access to cheap reduction opportunities. Its overall performance, however, is unlikely to be large (Ellis et al., 2004). The CDM is impeded by substantive transaction costs, resulting from the need to assess each project, prove it is additional to what would have happened otherwise, and to define an appropriate baseline. Relaxing the additionality criteria may augment neither the efficacy of the CDM nor its possible benefits for developing countries (Asuka and Takeuchi, 2004). As a result, many analysts believe that the CDM, as currently designed, may only play a minor role – though, arguably, this also results from a weaker demand for credits following the withdrawal of the USA from the protocol.

Another difficulty is that the CDM is unlikely to be effective against leakage. An agreement effective against this would need to create an opportunity cost for all emissions wherever they take place. This would be possible with a frictionless project-based mechanism if the baselines against which to credit emission reductions were comparable in both industrialized and developing countries. This is not what was decided in the Marrakech Accords. An efficient plant could possibly be closed in the industrialized world as a result of a carbon constraint, and its production replaced by a less efficient plant in a developing country, creating leakage. The CDM would not prevent this happening. It may even give such leakage some additional incentive if a newly-built plant is more efficient than those in the host country serving as reference for the baseline, and could thus earn some credits.

Sector-wide CDM, as suggested by Samaniego and Figueres (2002) and Chung (2003), or more broadly sector crediting mechanisms (Bosi and Ellis, 2005), could probably reduce transaction costs. It remains to be seen, however, whether this concept could alleviate concerns about emission leakage and competitiveness. Given these, moving into plain emissions trading with developing countries may be necessary for industrialized countries to adopt more ambitious commitments in the future. A global regime would avoid emissions leakage even if some countries were allocated surplus emissions beyond their needs, since greenhouse gas emissions would have the same opportunity cost everywhere. Any additional emission in such countries would represent a lost opportunity to sell. This loss entails the same cost as buying the permits to cover this emission in a constrained country. However, developing countries are unlikely to accept soon the type of fixed and binding quantified objectives that characterize the Kyoto Protocol.

4.5.3. The dynamic target, non-binding target options

Several options have been suggested to pull developing countries into global emissions trading regimes. The most common are dynamic targets and non-binding targets.

Dynamic targets (or 'indexed targets') are indexed according to an agreed variable, for example on the actual economic growth. Assigned amounts would be adopted in advance and based on some expectation – in this example relative to GDP growth, at a country level, or to some metrics of output, at the entity level. Then, if the economic growth were more or less than expected, these assigned amounts would be revised upward or downward. Dynamic targets would thus reduce the cost uncertainty that stems from uncertain emission trends – i.e. the uncertainty on the levels of emission cuts needed to reach a fixed target. Dynamic targets, however, are much less likely to address uncertainty about the costs of future abatement options or technologies. Dynamic targets could, in principle, be an option for both developed and developing countries, since they allow for full differentiation – either through varying assigned amounts or through indexation formulas (IEA, 2002; for a more recent discussion, see Philibert and Reinaud, 2004).

Intensity targets (defined as a ratio of greenhouse gas emissions to GDP) represent a particular form of dynamic targets. Ellerman and Wing (2003) suggest a simple and general formula for a 'growth-indexed emission limit' that combines a fixed and an intensity target. The degree of indexing, i.e. the relative weights of the two opposite forms, can take any value between 0 (pure fixed targets) and 1 (pure intensity targets).

Another option for developing countries would be *non-binding targets* (or 'no-lose targets'). These targets may provide – through emissions trading – an incentive for emission reductions, where sales could occur if (and only if) actual emissions are less than the targets (Philibert, 2000). The existence of such an incentive, however, requires that other countries are potential buyers bound by firm targets.

There are different ways to ensure that countries with non-binding targets only sell emission allowances that exceed the coverage of their actual emissions. The most effective may be to require countries that have over-sold to purchase enough allowances to cover their actual emissions up to the level of the non-binding target – but not beyond (Philibert and Pershing, 2001). A commitment period reserve, similar to that instituted by the Marrakech Accords, would also limit inadvertent mistakes.

Non-binding targets are progressively gaining support, or at least interest, from various experts from industrialized countries (e.g. Bodansky, 2003), newly industrialized ones (e.g. Chan-Woo, 2002), or developing countries such as India (e.g. Dasgupta and Kelkar, 2003) or China (e.g. Chen, 2003). Non-binding targets might be fixed or dynamic, country-wide or sector-wide. Dynamic non-binding targets would offer developing countries a greater chance to participate in international emissions trading despite possible economic surprises.

Apart from alleviating concerns about emission leakage, plain emissions trading could possibly offer a sharp reduction in transaction costs compared with project-based mechanisms, since the baseline would be established once and for all (at least for a 'commitment period') for a whole sector or country. This, however, would have a cost in terms of institutional requirements, as noted, in particular, by Baumert et al. (2003).

5. Conclusions

Effective national policies against climate change should comprehend both technology push and technology pull policies, mobilize short-term and long-term potential, and build upon learning-by-doing processes. They would probably use a wide number of instruments either to supplement broad market-based policy instruments or make them more efficient by specifically dealing with various market failures.

As suggested in this article, similar conclusions might be relevant in the area of international negotiations. Increased international technology collaboration would help especially with respect to research and development, and a variety of tools can be conceived to speed dissemination of innovative climate-friendly technologies in developing countries. More comprehensive frameworks giving directly or indirectly a price to carbon emissions might remain a first choice. Establishing a broader trading regime deserves constant efforts and innovative solutions to overcome the difficulties experienced thus far and alleviate the concerns that have been expressed by developing countries.

Acknowledgements

The views expressed here are those of the author and do not necessarily reflect the views of the IEA, its Secretariat or its Member States.

References

Asuka, J., Takeuchi, K., 2004. Additionality reconsidered: lax criteria may not benefit developing countries. Climate Policy 4, 177–192.

Barrett, S., 2003. Environment and Statecraft. Oxford University Press, Oxford.

Baumert, K., Perkaus, J.F., Kete, N., 2003. Great expectations: can international emissions trading deliver an equitable climate regime? Climate Policy 3(2), 107–187.

Benedick, R.E., 2001. Striking a new deal on climate change. Issues in Science and Technology (online), Fall [available at http://www.issues.org/issues/18.1/benedick.html].

Bodansky, D., 2003. Climate commitments: assessing the options. In: Aldy, J.E. et al. (Eds), Beyond Kyoto: Advancing the International Effort Against Climate Change. Pew Center on Global Climate Change, Arlington, VA, pp. 37–60.

Bosni, M., Ellis, J., 2005. Exploring options for sectoral crediting mechanisms. IEA/OECD Information paper, Paris.

Buchner, B., Carraro, C., Cersosimo, I., Marchiori, C., 2003. Back to Kyoto? US Participation and the Linkage between R&D and Climate Cooperation. Fondazione Eni Enrico Mattei, Venice, Italy.

Chan-Woo, K., 2002. Negotiations on climate change: Debates on commitment of developing countries and possible responses. East Asia Review 14: 42–60.

Charnovitz, S., 2003. Trade and Climate: Potential Conflicts and Synergies, in Aldy et al., Beyond Kyoto – Advancing the international effort against climate change. Pew Center on Global Climate Change, Arlington, VA.

Chen, Y., 2003. Chinese perspectives on beyond-2012. Presentation at the open symposium, International Climate Regime Beyond 2012: Issues and Challenges, October 7, Tokyo, Japan.

Chung, R., 2003. CDM-linked voluntary dynamic target. Presentation at the CCAP's Dialogues on Future Actions and Clean Development Mechanism, October, Jeju Island, South Korea.

Dasgupta, C., Kelkar, U., 2003. Indian perspectives on beyond-2012. Presentation at the open symposium, International Climate Regime Beyond 2012: Issues and Challenges, October 7, Tokyo, Japan.

Edmonds, J.A., 1999. Beyond Kyoto: toward a technology greenhouse strategy. Consequences 5(1), 17–28 [available at http://www.gcrio.org/CONSEQUENCES/vol5no1/beyond.html].

Edmonds, J.A., 2002. Atmospheric stabilization: technology needs, opportunities, and timing. In: Kennedy, D., Riggs, J.A. (Eds), U.S. Policy and the Global Environment: Memos to the President. The Aspen Institute, Aspen, CO, pp. 46–71.

Edmonds, J.A., Wise, M., 1999. Exploring a technology strategy for stabilising atmospheric CO_2. In: Carraro, C. (Ed.), International Environmental Agreements on Climate Change. Kluwer Academic Publishers, Dordrecht, The Netherlands, pp. 131–154.

Ellerman, D.A., Wing, I.S., 2003. Absolute versus intensity-based emission caps. Climate Policy, 3 [Supplement 2], S7–S20.

Ellis, J., Winkler, H., Morlot, J., 2004. Taking stock of progress under the CDM. OECD/IEA Information Paper, COM/ENV/EPOC/IEA/SLT(2004)4.

Gagnon-Lebrun, F., 2004. Cooperation in agriculture: R&D on high-yielding crop varieties. International Technology Cooperation Case Study 2. OECD/IEA Information Paper, COM/ENV/EPOC/IEA/SLT(2004)9.

Grubb, M., Hope, C., Fouquet, R., 2002. Climatic implications of the Kyoto Protocol: the contribution of international spillover. Climatic Change 54, 11–28.

Heller, T.C., Shukla, P.R., 2003. Development and climate: engaging developing countries. In: Aldy, J.E. et al. (Eds), Beyond Kyoto: Advancing the International Effort Against Climate Change. Pew Center on Global Climate Change, Arlington, VA, pp. 111–140.

Hoffert, M.I., Caldeira, K., Benford, G., Criswell, D.R., Green, C., Herzog, H., Jain, A.K., Kheshgi, H.S., Lackner, K.S., Lewis, J.S., Lightfoot, H.D., Manheimer, W., Mankins, J.C., Mauel, M.E., Perkins, L.J., Schlesinger, M.E., Volk, T., Wigley, T.M.L., 2002. Advanced technology paths to global climate stability: energy for a greenhouse planet. Nature 298, 981–987.

Hourcade, J.-C., Shukla, P., 2001. Global, regional, and national costs and ancillary benefits of mitigation. In: Metz, B., Davidson, O., Swart, R., Pan, J. (Eds), Climate Change 2001: Mitigation. Contribution of Working Group III to the Third Assessment Report of the Intergovernmental Panel on Climate Change. Cambridge University Press, Cambridge, UK, Ch. 8.

IEA, 2000. Experience Curves for Energy Technology Policy. OECD/IEA, Paris.

IEA, 2002. Beyond Kyoto: Energy Dynamics and Climate Stabilisation. OECD/IEA, Paris.

IEA, 2003. Cool Appliances: Policy Strategies for Energy-Efficient Homes. OECD/IEA, Paris.

Martin, J.-M., 1988. L' intensité énergétique de l' activité économique dans les pays industrialisés: Les évolutions de très longue période livrent-elles des renseignements utiles? Économies et sociétés (4) Avril.

Maurer, C., Bhandari, R., 2000. The Climate of Export Credit Agencies. WRI Climate Notes, World Resource Institute, Washington DC.

Moomaw, W., Moreira, J., 2001. Technological and economic potential of greenhouse gas emissions reduction. In: Metz, B., Davidson, O., Swart, R., Pan, J. (Eds), Climate Change 2001: Mitigation. Contribution of Working Group III to the Third Assessment Report of the Intergovernmental Panel on Climate Change. Cambridge University Press, Cambridge, UK, Ch. 3.

Pacala, S., Socolow, R., 2004. Stabilization wedges: solving the climate problem for the next 50 years with current technologies. Science 305, 968–972.

Philibert, C., 2000. How could emissions trading benefit developing countries? Energy Policy 28, 947–956.

Philibert, C., 2003. Technology, innovation, development and diffusion. OECD/IEA Information Paper. COM/ENV/EPOC/IEA/SLT(2003)4.

Philibert, C., 2004a. International technology cooperation and climate change mitigation. OECD/IEA Information Paper. COM/ENV/EPOC/IEA/SLT(2004)1.

Philibert, C., 2004b. International technology collaboration case study: concentrating solar power technologies. OECD and IEA Information Paper. COM/ENV/EPOC/IEA/SLT(2004)8.

Philibert, C., 2005. Energy demand, energy technologies, and climate stabilisation. In: Proceedings of the IPCC Expert Meeting on Industrial Technology Development, Transfer and Diffusion, held in Tokyo, 21–23 September 2004.

Philibert, C., Pershing, J., 2001. Considering the options: climate targets for all countries. Climate Policy 1, 211–227.

Philibert, C., Podkanski, J., 2005. International technology collaboration case study: clean coal technologies. OECD/IEA Information Paper. Com/ENV/EPOC/IEA/SLT(2005)4.

Philibert, C., Reinaud, J., 2004. Emissions trading: taking stock and looking forward. OECD and IEA Information Paper, Paris.

Samaniego, J., Figueres, C., 2002. Evolving to a sector-based clean development mechanism. In: Baumert, K. (Ed.), Options for Protecting the Climate. World Resource Institute, Washington, DC.

Sharma, A., 2000. Whose carbon hypocrisy? Should Northern groups be pushing international financial institutions to stop funding fossil fuels projects in the South? Down to Earth 19(10), October 15 [see http://www.downtoearth.org.in/more_stories.asp?currentpage=85&news_id=20788].

Sussman, F., 2003. Harnessing financial flows from export credit agencies for climate protection. Prepared for the expert dialogue meeting on Future International Actions to Address Global Climate, October 20–23, Jeju Island, Korea.

Tebar Less, C., 2003. Overview of work on the transfer of environmentally friendly technologies. COM/ENV/TD(2003)32, OECD, Paris.

Climate Policy 5 (2005) 309–328

Post-Kyoto climate policy targets: costs and competitiveness implications

Christian Azar*

Physical Resource Theory, Chalmers University, 412 96 Göteborg, Sweden

Received 17 January 2005; received in revised form 11 March 2005; accepted 24 March 2005

Abstract

This article starts with a review of climate policy targets (temperature, concentrations and emissions for individual regions as well as the world as a whole). A 20–40% reduction target for the EU is proposed for the period 2000–2020. It then looks at costs to meet such targets, and concludes that there is widespread agreement amongst macro-economic studies that stringent carbon controls are compatible with a significant increase in global and regional economic welfare. The difference in growth rates is found to be less than 0.05% per year. Nevertheless, concern still remains about the distribution of costs. If abatement policies are introduced in one or a few regions without similar climate policies being introduced in the rest of the world, some energy-intensive industries may lose competitiveness, and production may be relocated to other countries. Policies to protect these industries have for that reason been proposed (in order to protect jobs, to avoid strong actors lobbying against the climate policies, and to avoid carbon leakage). The article offers an overview of the advantages and drawbacks of such protective policies.

Keywords: Atmospheric stabilization; CO_2; Costs; Post-Kyoto targets; Competitiveness

1. Introduction

The United Nations Framework Convention on Climate Change (UN, 1992) calls for a 'stabilization of greenhouse gas (GHG) concentrations in the atmosphere at a level that would prevent dangerous anthropogenic interference with the climate system'. This ultimate objective of the climate convention forms the backbone of international climate politics. It calls upon us to act so as to make sure we do not cause unacceptable damage to humans, human societies, and ecosystems. Several key questions emerge:

- What level of climate change is dangerous? How does that translate into a concentration target for atmospheric greenhouse gases, and ultimately emission targets in the near, medium and long term?

* Corresponding author. Tel.: +46-31-7723132
E-mail address: frtca@fy.chalmers.se

- What are the costs of meeting those targets?
- How is the competitiveness of one region affected by policies that would deliver such emission reductions, if other regions do not adopt similar climate policies? What policy measures are available to address these concerns, and how do they work?

This article was initially prepared for a EFIEA workshop on EU strategies on post-2012 climate change policies with EU climate negotiators in Scheveningen, Holland on 30–31August 2004, where I was asked to address these questions. Clearly, a detailed review of the literature on these broad and admittedly varying topics cannot possibly be offered within the space of a single article. Therefore I have attempted to offer a review of key points that have emerged in the literature, interspersed with some personal viewpoints.

The article starts off with a discussion of dangerous anthropogenic interference with the climate system, and moves on to review and propose emission reduction targets required to meet a 2°C target (see Section 2). The costs to meet these targets are assessed in Section 3 and it is concluded that the burden sharing of the costs, rather than the total cost as such, is likely to be the most important obstacle to more ambitious climate policies. For that reason, Section 4 addresses the concerns about loss of competitiveness and possible ways to deal with it. Some conclusions are offered in a final section.

2. Climate policy targets

A precise statement of what constitutes 'dangerous anthropogenic interference' is not possible, since (a) the degree of harm from any level of climate change is subject to a variety of uncertainties and (b) the extent to which any level of risk is 'acceptable' or 'dangerous' is a value judgment (Azar and Rodhe, 1997; Schneider et al., 2000). Science can provide estimates about expected climatic changes and associated ecological and societal impacts, but ultimately the question of what constitutes dangerous has to be settled in the political arena – given, of course, the best scientific assessments available about the likelihood of various potential outcomes.

Several authors have focused on thresholds in the climate system, beyond which large-scale, often irreversible, changes take place (see Rial et al., 2004, for an overview of non-linearities, feedbacks and critical thresholds in the climate system, and Hulme, 2003, for a discussion about how human societies may cope with such changes). Examples of such thresholds include a shut-down of the thermohaline circulation, disintegration of the West Antarctic ice sheet, disintegration of the Greenland ice sheet, widespread bleaching of coral reefs, and disruption of other ecosystems (see Schneider and Lane, 2005, for a summary of temperature thresholds for each of these impacts).

The European Union (EU, 2005) as well as several scientists and research groups (e.g. Rijsberman and Swart, 1990; the Scientific Advisory Council on Global Change to the Federal Government of Germany (WBGU), 1995; Alcamo and Kreileman, 1996; Azar and Rodhe, 1997; Grassl et al., 2003; the International Climate Change Taskforce (ICCT), 2005) have argued in favour of an upper limit on the increase in the global annual average surface temperature set at or around 2°C above pre-industrial temperature levels.

Other researchers have analysed the data and proposed similar targets. O'Neill and Oppenheimer (2002) conclude that a 1°C target (above 1990 levels) may be required in order to prevent severe damage to coral reefs, a 2–3°C target to protect the West Antarctic ice sheet and a 3°C target to protect the thermohaline circulation. Arnell et al. (2002) find that stabilization at 550 ppm CO_2

'appears to be necessary to avoid or significantly reduce most of the projected impacts in the unmitigated case' (in their 550 ppm CO_2 run, the global mean temperature more-or-less stabilizes at about 2°C above 1990 levels by the year 2200).

Hare (2003) points out that certain ecosystems (in the arctics or in alpine environments, and coral reefs) may also be severely damaged at global temperature increases below 2°C. Hansen (2005) argues in favour of a temperature increase at a maximum of 1°C above current temperatures, based largely on concerns about the risk for rapid disintegration of the Greenland and West Antarctic ice sheets. Oppenheimer and Alley (2004, 2005) offer an insightful assessment of the role of the possible melting of ice sheets in determining dangerous anthropogenic interference.

Mastrandrea and Schneider (2004) and Wigley (2004) have developed subjective probability density functions for the temperature level at which dangerous anthropogenic interference takes place, based on the so-called 'burning embers' diagram of the IPCC (2001b, see ch. 19). Their median estimates lie at 2.8°C and 3°C, respectively.

Clearly, one should be careful to interpret thresholds as very sharp tipping points beyond which damages suddenly become dangerous or unacceptable for humanity as a whole. Carlo Jaeger (cited at http://www.realclimate.org/index.php?p=115) has argued that setting such a limit is nevertheless sensible, since it is a way to collectively deal with risks. He has made the analogy with setting speed limits: When we set a speed limit at 90 km/h, there is no 'critical threshold' there – nothing terrible happens if you go to 95 or 100 km/h. But, at some speed, risks (the number of accidents and the impacts) would exceed acceptable levels.

Finally, this discussion should not be understood as a call for governments to initiate formal negotiations on long-term temperature targets that should be adhered to over the next hundred years. Such negotiations are likely to end up in a nightmare of complexities and problems. Perhaps even more importantly, uncertainty about the climate system, impacts, costs, baseline emissions, and so on, suggest that adhering to one target over such a long time period would not be very wise. Rather, the purpose of endorsing a target, or merely thinking about a target, is that it gives guidance as to what may be required during the next couple of decades in order to make sure that we do not act now so that we get locked into a future with unacceptable climate damages.

2.1. Temperature and concentration

Here, I pursue the view that a global annual average surface temperature increase of more than 2°C above pre-industrial levels should be avoided (in line with ambitions expressed by the European Union) and estimate the required concentration and emission targets.

In Figure 1, the relation between atmospheric concentrations and the global *equilibrium* average annual surface temperature change is shown (see Azar and Rodhe, 1997). In the graph, the climate sensitivity (the equilibrium temperature change for a doubling of pre-industrial CO_2 concentrations) is assumed to be 1.5–4.5°C/CO_2-equivalent doubling (IPCC, 2001a; Kerr, 2004). Further, a net contribution to the radiative forcing from other greenhouse gases and aerosols of 1 W/m^2 is assumed.[1]

It can be seen from Figure 1 that a CO_2 concentration of 550 ppm is expected to lead to a temperature increase in the range 1.9–5.5°C. For 350 and 450 ppm CO_2, the expected equilibrium temperatures are 0.9–2.6°C and 1.4–4.5°C, respectively. Thus, in order to be relatively certain that a 2°C target is actually met, CO_2 concentrations would have to remain below 400 ppm.

There is a growing literature aiming at developing probability density functions for climate sensitivity (e.g. Andronova and Schlesinger, 2001; Forest et al., 2001; Wigley and Raper, 2001;

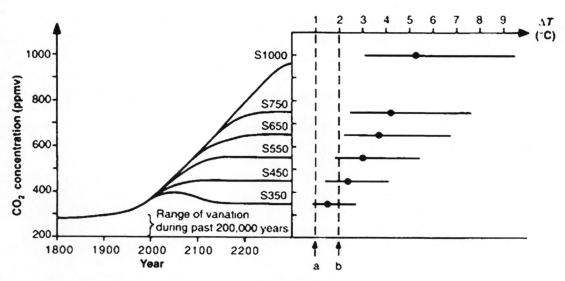

Source: Azar and Rodhe (1997), reprinted with permission from *Science*.

Figure 1. Global average surface equilibrium temperature change for various stabilization targets. Dashed line (a) refers to an estimate of the maximum natural variability of the global annual average surface temperature over the past millennium, and dashed line (b) shows the 2°C temperature target.

Gregory et al., 2002; Stainforth et al., 2005). These studies support the IPCC range in general, but have tails below 1.5°C and higher than 4.5°C; in some cases much higher.[2]

Taking these distributions into account would make it possible to estimate probabilities for the level below which the concentration of CO_2 has to stay in order to avoid any given temperature increase. Such studies have been performed by Baer (2004) and Meinshausen (2005), who both conclude that 400 ppm CO_2-equivalent (corresponding approximately to 360 ppm CO_2 only) is probably required if we are to be relatively certain to avoid a temperature increase of 2°C.

2.2. Global emission trajectories towards 2°C target

In Figure 2, emission trajectories towards 350, 450 and 550 ppm are shown. All these concentration targets are potentially compatible with a 2°C temperature target but with very low probabilities for the 550 ppm case (as illustrated in Figure 1). It can be seen that the implications for the global energy system over the next 50 years differ radically depending on the climate sensitivity. If the climate sensitivity is so low that the 550 ppm CO_2 case is compatible with the 2°C target, then global carbon emissions may increase by 20% until the year 2050. On the other hand, if the climate sensitivity is so high that the 350 ppm concentration target is required, then emissions need to be reduced by 75% over the next 50 years. The importance of the climate sensitivity for the required emission trajectory towards a 2°C target has also been highlighted by Caldeira et al. (2003).

A key question is what this uncertainty about the climate sensitivity and the ultimate temperature target implies for the near-term emission reduction requirements. This question received widespread attention with the publication by Wigley et al. (1996), who argued that delaying emission reductions compared to the IPCC stabilization scenarios (IPCC, 1994), would not only be possible but also more cost-efficient.

Carbon emission trajectories towards 350, 450, 550 ppm

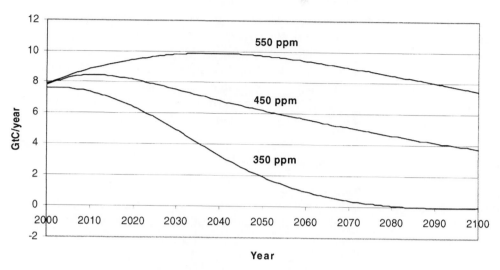

Figure 2. Emission pathways towards 350, 450 and 550 ppm developed as the average of IPCC S350–S550 scenarios (IPCC, 1994; Wigley et al., 1996). Each pathway may be compatible with a 2°C temperature target, but this would require a climate sensitivity of around 1.5°C/CO_2-equivalent doubling for the 550 ppm CO_2 target.

But the challenge now, as IPCC (1996) states, 'is not to find the best policy today for the next hundred years, but to select a prudent strategy and to adjust it over time in the light of new information'. If we follow an emission trajectory towards say 550 ppm, and later on find out that a 400 ppm target is required, the long-life time of carbon in the atmosphere and the inertia of energy capital as well as the political system, may make it impossible to meet this lower target (see, e.g., Ha-Duong et al., 1997; Schneider and Azar, 2001).

Azar and Rodhe (1997) concluded that 'until it has been proven that a temperature increase above 2°C is safe or that the climate sensitivity is lower than the central estimate, the projections shown ... suggest that the global community should initiate policies that make stabilization in the range 350 to 400 ppmv possible'.

It is in this context interesting to reflect on the policy implications of a recent publication by Wigley (2004). He assumes a probability density function for the temperature target, with a mean at 3°C, and combines that with a probability density function for the climate sensitivity (Wigley and Raper, 2001). Given his mean target of 3°C, he finds that there is a 50% probability that the concentration of CO_2 needs to be stabilized below 536 ppm. However, he also finds that there is a 23% probability that the concentration of CO_2 needs to be stabilized below 400 ppm CO_2. Thus analyses with medium targets of 3°C and 536 ppm CO_2 could also well justify decisions to act now so as to keep 400 ppm CO_2 within reach.

The exact reduction target in the near term that these considerations imply depends on whether one allows for a temporary overshoot of the concentration or the temperature target. For instance, if negative carbon emissions can be obtained (through the use of air capture or biomass with carbon capture and storage; see Lackner, 2003; Obersteiner et al., 2001), then a 350 ppm concentration target by the year 2100 could be met even if atmospheric concentrations exceed

400 ppm by the middle of the century (see, e.g., Azar et al., 2005). An overshoot of the temperature target might lead to irreversible changes in the climate system or in ecosystems, which means that the pathway to the target is of importance. Allowing for such temporary overshoots might, thus, come into conflict with the recognition in the UNFCCC that it is not only the absolute level of climatic change, but also the rates of change, that matter. Another factor determining how much needs to be done in the near term is the inertia in the energy system and the political system. If the maximum rate with which emissions may be reduced is assessed to be low, then relatively more ambitious policies need to be introduced in the near term (see Meinshausen, 2005, for illustrations of what delayed abatement implies for subsequent required rates of change).

2.3. Regional emission targets

Breaking down global emission pathways into reduction targets for individual countries or regions is probably one of the more contentious challenges for climate negotiators. It should be clear that there is no single correct answer to the question of how much the EU needs to reduce the emissions in order to meet a, say, 450 ppm concentration target. The reason for that is not only that there is some degree of freedom as to when the reductions should take place, as discussed above, but also – and perhaps more importantly – that there are several different methods that can be used to share the burden of emission reductions between countries and regions; e.g. equal per capita, contraction and convergence (Meyer, 2000), multistage, intensity targets, global triptych and multi-sector convergence (see, e.g., den Elzen, 2002; Grassl et al., 2003; Höhne, 2005).

Due to space limitations, it is not possible to review these results in detail. Instead, I will offer an illustration of the implications of one approach – contraction and convergence by the year 2050 with a focus on CO_2 for three different concentration targets (350, 450 and 550 ppm). Results where other approaches are taken and when all the Kyoto gases are considered are discussed later.

In Figure 3, per-capita emissions in the European Union and China over the next 50 years that would be compatible with a global effort to meet these three targets are shown. The emission pathways are developed in the following way. It is assumed that all countries receive emissions allowances for the year 2000 that represent their current emissions. For the year 2050, allowances are allocated on a per-capita basis globally. For the years in between, a linear weighting scheme is assumed.[3] In addition, I have assumed that the contributions from deforestation and land-use changes drop linearly from 1.5 GtC/year at present to zero by the year 2050. The global population reaches 9.1 billion by the year 2050 (UN, 2004).

For the year 2050, the required reduction in EU lies in the range 50% (for a 550 ppm target) to 90% (350 ppm). It is worthwhile to note that there is such a sharp reduction requirement for the 550 ppm target despite the fact that the global carbon emission trajectory leading to 550 ppm actually increases by 20% (see Figure 2). The reason for this is that the contraction and convergence approach requires that emission allowances should be allocated on a per-capita basis.

For the year 2020, the per-capita reduction targets for the EU, should be in the range minus 20–40% compared to the year 2000 for the 350 and 450 ppm targets, respectively. I am deliberately rounding numbers in order to avoid creating the impression that one can be very precise in establishing what needs to be done in one region in the near term in order to meet a global long-run target. It is interesting to compare these targets with those proposed by the Council of the European Union (on 10 March 2005). The EU proposed that the developed countries adopt reduction targets (for all Kyoto greenhouse gases) in the order of 15–30% below 1990 by the year 2020 (see EU, 2005).

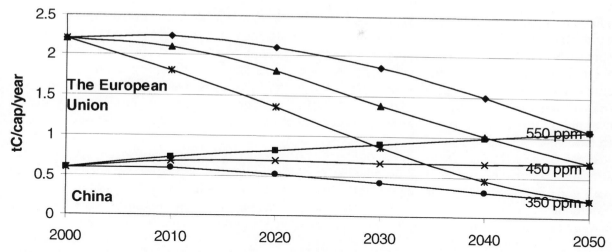

Figure 3. Per-capita emission trajectories for China and the EU towards 350, 450 and 550 ppm, under contraction and convergence by 2050. Population scenarios are taken from UN (2004) and per-capita emissions for the year 2000 from Marland et al. (2003).

Other, more detailed assessments of the reduction requirements generally fall in this range, not only for the contraction and convergence but also for other allocation methods; e.g. the Triptych regime and various forms of multistage models (see den Elzen, 2002; Nakicenovic and Riahi, 2003; den Elzen et al., 2005; Höhne, 2005; Persson et al., 2005).

den Elzen and Berk (2004), for instance, find that a reduction of all Kyoto greenhouse gases by approximately 30% is required over the years 1990–2025 in an 'enlarged EU' in order to meet a 550 ppm CO_2 equivalent target for not only contraction and convergence by 2050 but also for Triptych and for a multistage approach. The reason why their number is lower than the upper range in our estimate is that our higher value reflects a more ambitious reduction target (compatible with 350 ppm CO_2).

Cases where the allocation approach does have a significant impact on the near-term reduction requirements include (rather obviously) equal per-capita now, contraction and convergence by the year 2100, which gives less stringent reductions in the North (and correspondingly more stringent targets in the South), and the Brazilian proposal, which requires somewhat steeper reductions in the Annex-1 countries because of its focus on historical responsibility.

For China the large difference in the 350 and 550 ppm global emission trajectory (Figure 3) translates into either a possibility to increase its per-capita emissions by 80% (in the 550 ppm case) or decrease them by 70% in the 350 ppm case.

I chose to include only the EU and China in the graph in order not to complicate the picture with too many regions, but it is worthwhile to note that the results for the EU also hold (in broad terms) for Japan, the Former Soviet Union FSU) and South Africa. The USA, Canada, Saudi Arabia and Australia have substantially higher per-capita emissions, so the reduction requirements are sharper. The results for China hold roughly also for fossil-fuel-related emissions from Latin America. India, Africa and Indonesia emit roughly half as much per capita as China and Latin America and may thus be allowed to increase their emissions of CO_2. On the other hand, methane and N_2O emissions in India, Indonesia and southern Africa are larger than the emissions of fossil carbon, so taking these gases into account implies more stringent emission targets for these countries.[4]

It may also be noted that there are many countries that traditionally refer to themselves as belonging to the South, that emit more or much more than 1 tC/cap/year (e.g. Malaysia, Iran, South Korea, Mexico, Argentina and, as already mentioned, Saudi Arabia and South Africa).

Different allocation methods yield more variable results for developing countries than for developed countries, in particular for countries with very low emissions at present. For India and sub-Saharan Africa the choice of methods may imply differences in emission profiles (or allocated allowances) that amount to several hundred percent of their current per-capita emissions (see, e.g., Höhne, 2005, ch. 6, figs 4 & 6).

Finally, the actual emissions under a contraction and convergence approach, or any other allocation approach, will depend on whether trade in allowances is allowed or not. Analyses of such trade in allowances are uncertain, since they depend on assumptions about baseline economic development, options to reduce emissions in different regions, political pressure to carry out most of the reductions domestically, etc. For examples of such studies, see Nakicenovic and Riahi (2003), den Elzen et al. (2005) and Persson et al. (2005). Most studies conclude that rich countries generally end up being buyers of permits under a contraction and convergence approach by the year 2050 scheme aiming at 450 ppm, but that China also rather soon ends up being a net buyer (because of its high growth rate and large coal resources).

3. Overall cost of mitigation

There is much concern about the cost of meeting stringent climate targets. In the public debate, claims are even made that climate policies will threaten our current standard of living. But what does the economics literature tell us? In the latest IPCC assessment, the cost of stabilizing the atmospheric concentration of CO_2 at 450, 550 and 650 ppm is estimated to lie in the range US$2.5–18 trillion, US$1–8 trillion and roughly US$0.5–2 trillion, respectively (IPCC, 2001a, ch. 8).

In order to better understand what these numbers mean, it may be useful to view them in light of the expected overall global economic development. This is done in Figure 4 (see Azar and Schneider, 2002). The difference between global income under a 350 ppm scenario and the business-as-usual income (a growth rate of 2.1% per year) represents a net present value cost of US$18 trillion. Thus, although trillion-dollar costs are large in absolute terms, they are minor compared with the perhaps ten-fold increase in global income expected over the next hundred years. Similar observations can be made for the cost of meeting near- and mid-term climate targets.

Figure 4 should not be interpreted as if we were trying to argue that it is inexpensive to meet low stabilization targets. The point is to reject the rather widespread misperception that climate policies are not compatible with continued economic development. If policymakers and the general public would understand that the cost amounts to a few years' delay in becoming ten times richer by the year 2100 or as a difference in growth rate of on average less than 0.05% per year – hardly noticeable even in retrospective! – the willingness to accept climate policies would probably be higher.

It would also be wrong to conclude that the minor difference in growth rates between a stringent climate policy and business-as-usual implies that the low carbon future will materialize by itself. On the contrary, major efforts are required to achieve the almost complete transformation of the energy system that is required (see IPCC, 1996b, ch.11, or Azar et al., 2003, for examples of energy scenarios meeting stringent climate targets). There is in particular a need for (i) introducing and continually increasing the cost of emitting CO_2 (through the use of a tax, or a cap-and-trade system), (ii) for standards for energy efficiency improvements, and (iii) for a concerted effort to

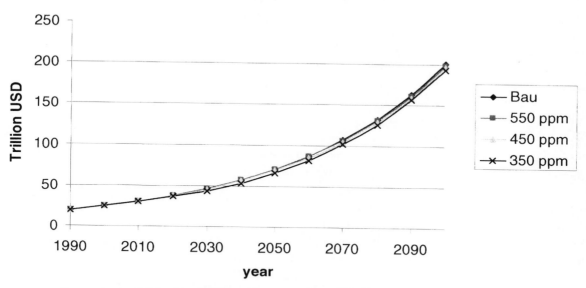

Source: Azar and Schneider (2002), reprinted with permission from Elsevier.

Figure 4. The development of global income, with and without climate policies. Climate damages are not quantified and thus not included in the graph.

enhance technology development not only through more R&D spending but also through the creation of niche markets for the emerging more-advanced carbon-free energy technologies (see Sandén and Azar, 2005).

4. Some perspectives on climate policy and the implications on competitiveness

The difficulties in achieving agreements on climate policies stem from many factors: for instance the fact that costs of climate change and of emissions abatement will not be shared equally across countries, there is not enough public awareness and support from climate policies, there is a widespread misperception that the costs of dealing with climate change will threaten overall economic welfare levels, and the cost of the policies will fall on people living now, whereas benefits will accrue to future generations.

An additional key obstacle is opposition from sectors or industries which would be heavily affected by climate policies. This aspect becomes particularly relevant if the policy ambitions differ across countries.[5] Climate policies would then, it is often argued, lead to relocation of production, which could be costly in terms of premature closure of industrial facilities and losses of jobs, and lead to increases in carbon emissions in other countries (sometimes referred to as 'carbon leakage').

The mere expectation that such competitiveness losses may occur is sufficient to set strong interest groups in motion against climate policies. The most well-known example is probably the Byrd–Hagel resolution in the US Senate in 1997, which explicitly stated that the USA should not accept any outcome in Kyoto unless it mandated 'specific scheduled commitments to limit or reduce greenhouse gas emissions for Developing Country Parties within the same compliance period'. Competitiveness concerns also partially explain why the EU chose to grandfather permits and why countries have been very generous when it comes to the total amount of allowances allocated in the EU Emissions Trading Directive (Grubb et al., 2005).

Thus, it is worthwhile to better understand the concerns about competitiveness and what governments may possibly do about it. Whether they should introduce protective policies is a political question that will not be addressed here. Rather, I will review insights from the literature and offer perspectives on questions such as 'What are the consequences of protecting, or not protecting, sensitive industries?' and 'What are the pros and cons of different protective policies?'.

Loss of nationwide competitiveness – or?

It is misleading to speak of losses of competitiveness at the country level as a result of climate policies. In fact, nationwide competitiveness is not even a well-defined concept in economics (see Krugman, 1994; Babiker et al., 2003). Households and transportation do not 'compete' with their likes in other countries. Further, according to the theory on international trade, an economy should specialize more in producing the goods it is comparatively better at, regardless of whether it has an absolute advantage over its trading partners. Implementing a uniform carbon price will shift advantage from carbon-intensive industries toward less carbon-intensive industries (compared with trading partners that do not implement such policies).

At the micro level, however, competitiveness is a useful concept. A company could be said to be competitive if it can produce goods at or below the prevailing market price. Energy- and carbon-intensive industries that face competition from regions without climate policies may lose competitiveness if the cost of energy and carbon increases.

But it should also be recognized that most industries have low energy costs compared to their turn-over, and these may even gain competitiveness and increase output (this is a common result in computable general equilibrium models, which even suggest that output in manufacturing industries may increase; see, e.g., Bergman, 1996). The way this could operate at the international market is as follows: a drop in the exports (or an increased import) of energy and carbon-intensive goods would eventually lead to a slight depreciation of the exchange rate (*ceteris paribus*). This depreciation would improve the competitiveness of manufacturing industries whose lower production costs (in international currency) would outweigh the impact of higher energy prices.[6] Thus, although it is not correct to talk about nationwide losses of competitiveness, a slight depreciation of the currency nevertheless implies higher import prices; i.e. a slight loss in real income.

Competitiveness of energy-intensive industries

Energy- and carbon-intensive industries include steel, aluminium, chemicals (e.g. fertilizers), cement and refineries.[7] Producers of these products have limited opportunities to pass on increases in production costs to consumers since the price is often set by international markets (where producers do not face the same carbon price). Electricity generation from fossil fuels is clearly also energy- and carbon-intensive, but if there is no trade in electricity with non-abating regions, then electric utilities can obviously not lose competitiveness to producers in these regions.[8]

For many of these energy-intensive companies, competitiveness, measured as their production costs compared with those of competitors outside the climate abating regions, is at stake. Table 1 shows estimated increases in production costs for a US$10/tCO$_2$ tax on various energy-intensive industries (assuming constant production technology). The cost increase includes the tax on on-site emissions and the higher electricity prices that result from the carbon tax.

Table 1. The impact of a US$10/tCO$_2$ carbon tax on the production cost of energy-intensive products (Reinaud, 2004)*

Impact of a US$10/tCO$_2$ tax	Steel basic oxygen furnace	Steel (electric arc furnace)	Cement	Newsprint	Aluminium
Cost increase (%)	7.7%	1.5%	18.6%	3.9%	2.4%
Total cost increase (US$/t)	20.6	3.4	8.7	4.5	28.6

* Reinaud writes that these numbers are rough estimates of the *upper* boundary of the costs since they do not include options to lower carbon emissions or electricity use in these industries. In addition, the cost number refers to the average plant. Further, the author has chosen to use the *average* carbon emission factor (gC/kWh) for electricity generation in Europe when estimating the impact on the electricity price. But, if the emission factor of the *marginal* electricity source were to determine the impact on the electricity price, the cost increase for aluminium could be more than twice as high, since it is the change in electricity price that is the most crucial parameter for aluminium.

In the EU Emissions Trading Scheme (EU ETS) the increase in average production cost is much smaller because of the grandfathering of emission permits. Energy-intensive companies are basically given as many permits as they need for this first phase, 2005–2007, and may chose to 'consume' these permits in order to keep the price impact down. For that reason, Carbon Trust (2004) concludes that the EU ETS is not likely to pose any significant threat to energy-intensive industries in Europe, except possibly for aluminium industries, which will face a higher electricity price but will not receive any grandfathered permits. The impact on the aluminium sector thus depends on the extent to which the electric utilities are successful in passing on the opportunity costs of the permits to consumers.

This observation is similar to the conclusions drawn from studies about the relocation of industries facing unilateral regulations of other environmental problems, e.g. sulphur, emissions of metals etc. The general result from the literature on this issue is that it has proven difficult to demonstrate a strong case for such relocation (Jaffe et al., 1995; Persson, 2003; Cole, 2004). It would be premature, however, to conclude that this would be the case for stringent climate policies, since the costs of dealing with the CO$_2$ problem per unit of output in energy-intensive industries is significantly higher than the cost of dealing with many other environmental problems.

In the longer term, e.g. if the EU aims at reducing emissions by 20–40% by the year 2020, carbon prices might be several times higher than US$10/tCO$_2$. Bollen et al. (2004), for instance, estimate that a permit price of €58/tCO$_2$ by the year 2020 would reduce emissions in EU-25 by 31% compared with 1990. Although the authors also emphasize that there is a great deal of uncertainty about the exact value of the permit price, it is nevertheless likely that the required permit price will be in the tens of euros per tonne of CO$_2$ and such high permit prices would be likely to lead to severe competitiveness problems for energy-intensive industries from companies that do not face similar carbon penalties.

Higher cost of climate policies if industries are protected – or?
Economic assessments generally find that the cost of meeting a *domestic* carbon target typically increases if protection of sensitive industries takes place (see, e.g., Böhringer and Rutherford, 1997; Babiker et al., 2000, 2003; Bye and Nyborg, 2003).[9] For instance, Böhringer and Rutherford (1997) found that the cost of meeting a 30% reduction target for Germany would increase from 0.6% of its GDP to 0.8% of GDP if energy-intensive industries are protected. The fundamental reason for the expected increase in cost is that lowering the tax, or in general the effort to reduce the emissions, in

one sector, means that more costly options have to be employed in other sectors. However, let us assume that the aim of the unilateral climate policy is to meet a global emission target (defined as the sum of the domestic emissions plus the impact on the emissions in the rest of the world). Then the cost is typically lowered if some form of protection of heavy industries takes place (see, e.g., Bergman, 1996; Hoel, 1996; Böhringer and Rutherford, 1997).

These results are all obtained with the use of general equilibrium models, which typically are poor at capturing non-equilibrium effects, such as unemployment.[10] For that reason they may underestimate the social costs associated with rapid closures of large industries.[11] In addition, these models are rarely, if ever, run under the assumption that other countries will eventually also initiate carbon abatement policies. If they do, it could be argued that it would be economically inefficient to pursue a policy that leads to relocation of industries away from Europe if it is believed that these industries would be competitive in a near future with similar carbon constraints in the rest of the world. Such considerations could offer an argument in favour of temporary protection, but they also imply that the risk for relocation of industries is lower than what one may conclude from static analysis. Companies are, of course, aware of the fact that other countries may also introduce climate policies.

Losing or gaining markets?

Even if there is some risk that energy-intensive industries may relocate to other regions if Europe unilaterally pursues more ambitious climate policies, it should also be kept in mind that such policies would probably enhance the development of carbon-efficient technologies in Europe. This may be economically positive for Europe in the longer term, since it is most likely that other countries will eventually start to abate carbon. European industries may at that time gain a competitive advantage on these new markets. The Danish export of wind power is an example worth noting. This perspective is sometimes referred to as the Porter hypothesis (Porter and van der Linde, 1995).

Furthermore, technology development that leads to more efficient technologies in Europe (say in the automotive industry, in electric appliances etc.) may set the standard also in other countries regardless of their climate ambitions. This would, in turn, lead to reductions in carbon emissions in their countries, i.e. a reversed form of carbon leakage (see, e.g., Grubb et al., 2002).

Options for protection

There are several different policy options that may be employed to protect the energy intensive industries from climate policies, e.g.,

- Allocate carbon emission allowances freely on the basis of past emissions (grandfathering).
- Introduce so-called border tax adjustments (BTA), i.e. import taxes and export subsidies, that level the playing field with countries outside the carbon-abating region.
- Differentiate mitigation efforts between sectors (different carbon tax levels, full tax exemptions, trading schemes that only cover certain sectors, as is the case with EU ETS, etc.).
- Direct subsidies to compensate industries that lose competitiveness.

In common for all these options is the fact that there will be methodological problems in the implementation phase and that protective policies may come into conflict with basic ambitions of achieving free trade and non-distorted markets. There have already been complaints about unfair allocations to companies in different countries in the case of the EU ETS. Another problem is that there would be a risk that these protective policies would be self-reinforcing in the sense that industries, once protected, will continue to claim the right to protection even when the carbon-abating efforts of other countries increase. The coal subsidies in Germany are a case in point, where subsidies amount to €82,000 per job in 2001 (see press release from the Federal Environmental Agency in Germany; FEA, 2003). Yet another problem is that there is a risk that one introduces policies to protect industries which really do not need protection. This would be the case for energy-intensive industries which plan to remain in the country but manage to get subsidies by threatening that they would relocate unless some form of compensation is given. Another example could be firms that would move abroad regardless of the climate policy, but stay only to get subsidies (e.g. aluminium industries in search of low cost electricity options – which may be found in regions with large hydro resources compared to the electricity consumption).

A difference between these policies is that some lead to the protection of the continued operation (e.g. direct subsidies that match the extra cost faced by the industries), whereas others aim at protecting the interests of the capital owners (e.g. grandfathered permits that could be sold and generate revenues to the capital owners even if the plant were to be taken out of operation).

Below, we will discuss some of these protective policies in more detail.

Cap and trade with grandfathering of emissions allowances

Grandfathering permits, i.e. the free allocation of permits based on historical emissions, rather than auctioning (or the use of taxes), has several drawbacks or features worth paying attention to. First, grandfathering is expected to increase the cost of meeting any given target substantially (see IPCC, 2001a). The reason for that is that the loss of government revenues that a tax or auctioned permits would have generated could have been used to offset distortive taxes.

A second important feature is that grandfathering based on historic emissions fails to offer protection to *electricity*-intensive industries (e.g. aluminium smelting; see, e.g., Reinaud, 2004; Carbon Trust, 2004). This has already caused concern amongst electricity-intensive industries in Europe.[12] For that reason, complementary measures may nevertheless be needed, for instance direct subsidies to electricity-intensive industries that cannot pass on increases in production costs to consumers. Spain and Ireland have introduced legislation that prevents electric utilities from raising the price of electricity (Reinaud, 2004).

A third potential problem is that if grandfathering of emission permits becomes the norm in environmental policy, the incentive to be proactive and reduce emissions in advance of environmental policy breaks down.

Fourth, energy-intensive industries often argue in favour of grandfathering so as to ensure continued operation in the face of 'unfair' competition from regions without (similar) climate policies. However, whether grandfathering offers such protection or not depends on how allocation decisions are made in subsequent commitment periods, and whether firms are behaving as profit maximizers or not. Permits allocated based on past emissions can be seen as a one-time donation to the capital owners. Whether the firm would continue to operate or not would then depend on the relation between the expected profits from selling the permits and the expected profits of continued operations. The time span over which the permits are allocated are here an important factor that determines the relative profitability of closing versus continued operation. Regardless of whether the plant closes down or not, such a policy would offer effective incentives to reduce the emissions, at least as long as the updating of the allowances for subsequent periods does not depend on the emissions in the preceding period.

If emission allowances are continually updated based on the emissions in the preceding period, then there would be incentives to increase emissions so as to get more permits. If the commitment periods are short, it is rather unlikely that it would be profitable to close down the firm, and under these conditions the policy would look more like a subsidy. This would protect the firm from closing down but in its extreme version would imply that there would be no climate policy at all. It may be noted that the decision on how to update allowances for the next period in the EU ETS is yet to be taken, so this is not simply an academic observation.

Finally, grandfathering to industries that may pass on most of the opportunity cost of the permits to consumers may see their profits increase as a result of climate policies. The value of the permits allocated to coal-fired power plants may actually be of the same order of magnitude as the value of the entire plant.[13] The fact that a carbon policy might lead to increased profits for a carbon-intensive industry might be difficult to digest, at least in the perspective of the 'polluter pays' principle.

One possible compromise would be to employ selective and partial grandfathering: selective in the sense that auctioning would be the norm but with grandfathering for the energy-intensive sectors, and partial in the sense that the companies would at most be grandfathered to the extent that profit levels do not increase (see Kågeson, 2000). Goulder (2005) reports that only a small share of the allowances need to be grandfathered in order to maintain profit levels in the US economy; the exact level depends on how much of the price increase may be passed on to the consumers. He concludes that 'major stakeholders can be compensated without significantly increasing the overall policy costs'.

Border tax adjustments

An interesting, but also complicated and for some contentious, approach might be to introduce import taxes (and possibly export subsidies) on carbon-intensive products, to (and from) countries in which there is no carbon abatement policy. The import tax would have the benefit that it would be close to equivalent (for European consumers) to a tax on production in other countries intended for European markets, and the export subsidy should be set so as to level the playing field in regions outside Europe (and all the other regions that have taken on climate policies).

The introduction of such border tax adjustment would almost certainly lead to problems with WTO rules (National Board of Trade, 2004), but pursuing this approach would, in addition to its immediate climate benefits, have the benefit of sending a message to other countries, as well as people not directly involved, interested or engaged in climate affairs, that the EU takes the threat of climate change seriously. Clearly, any country that would take on commitments with similar carbon prices as those that prevail in the EU would automatically be exempted from border tax adjustments, and the EU may argue that any country that has a problem with these tariffs can simply join the climate treaty (see also Hoel, 1996).

One problem with this approach is that it is very difficult, if not impossible, to calculate the correct level of the tariff on all products – just imagine keeping track of the embodied carbon emissions in each product entering the EU. For that reason, the only reasonable approach would be to include only a few products, e.g. steel, aluminium, some other metals, and fertilizers etc. The tax could be set based on some form of benchmarking, for example the best available technology, so as to make sure not to discriminate against any foreign producer who is very efficient. But even this approach would not be free from problems. In the case of aluminium, the emissions associated with its production would depend very much on whether coal or hydro is the marginal electricity source, and that choice (or property of the electricity system) has nothing to do with the best available technology to produce aluminium. Thus, it will not be possible to completely avoid the problem of site-specific emission factors. There is also a borderline problem: if energy-intensive materials (e.g. steel) are faced with an import tax, then what about manufactured goods (e.g. car bodies, cars etc.).

Differentiated efforts – including the European transport sector in the EU ETS
The EU ETS scheme only includes emissions from large point sources. Calls have been made to include also other sectors, e.g. the transportation sector and residential sectors. This could be done by requiring that importers and refineries need to hold permits for emissions that will be generated by users of gasoline and fuel oil. But such a decision would have implications, as we will see, for the competitiveness of energy-intensive industries.

There are basically two arguments in favour of inclusion. First, it would improve cost-efficiency of European climate policies (equalize carbon prices across a wider range of emission sources). Secondly, since the current EU ETS only covers some 40% of the overall CO_2 emissions in the EU, it has proven difficult to relate the target for the trading sector to the overall Kyoto target for the EU. By claiming that emissions will be reduced substantially in the non-trading sectors, it has been possible for several countries to allow for generous, perhaps too generous, allocations in the trading sectors.

The key argument against including other sectors is that a sufficiently strong target to comply with the Kyoto targets would imply that one of these sectors, the transportation sector, which probably has the largest willingness and capacity to pay for permits, would drive the price of the allowances to levels that would be difficult to deal with for the energy-intensive industries. It is in this context worth observing that the Swedish carbon tax on households and transportation is currently around US$100/t$CO_2$, whereas the permit price in the EU ETS is, in February 2005, around €10/tCO_2. ('The transportation sector would buy all the permits', exclaimed a frustrated representative for an energy-intensive industry to me recently) Such prospects could make it politically very difficult to introduce a sufficiently stringent cap in the trading sector because of lobbying from energy-intensive industries. In addition, including the transportation sector under the cap means that other measures to reduce the emissions in this sector will not lead to lower emissions, because the overall cap is already set.[14] Thus, there is a risk that the overall abatement will become less stringent if these sectors are also covered in the EU ETS.

5. Summary and conclusions

In this article, targets for the global average annual surface temperature, atmospheric concentration of CO_2 and emission of CO_2 have been reviewed and proposed. It is concluded that the EU needs to reduce emissions by 20–40% by the year 2020 compared with the year 2000 if we want to stabilize atmospheric concentration of CO_2 in the range 350–450 ppm CO_2 and pursue an approach based on contraction and convergence by the year 2050. For many developing countries, per-capita emissions are already above the per-capita targets by the year 2050, in particular for targets lower than 450 ppm. For developing countries with low emissions per capita, there is still room for substantial increases in emissions.

The article then assessed the cost of stabilizing the atmosphere at these levels. It is found that models that are generally perceived as being pessimistic find that the costs are compatible with continued impressive growth in global GDP. The reduction in growth rates, averaged over the entire century, is less than 0.05% per year. A key conclusion is that overall costs to meet stringent climate targets do not seem to be large enough to explain the strong resistance to the introduction of climate policies. Instead, it is the fact that the reductions will create winners and losers that probably causes the most severe opposition. This problem is aggravated by the fact that all countries of the world do not move ahead with climate policies at the same speed, and they are not likely to do so in the near future either.

It is concluded, however, that energy-intensive industries are not likely to lose competitiveness to any large extent under the current first phase of the EU ETS. For stricter emission reduction targets, such as those envisaged above, many energy-intensive companies would most probably lose competitiveness under the assumption that there would be no climate policies in major producer countries. If the rest of the world follows the EU in its climate ambitions, which of course is necessary for the EU climate policies to be meaningful, there would not be any need to introduce protective policies. Under such conditions would it be economically more efficient if the full cost of carbon were to be reflected also in the price of energy-intensive goods, since that would lead to substitution away from these materials.[15]

But as long as there are large differences in the climate ambitions across countries, there will be discussions about unfair competition and carbon leakage. Two contrasting positions may be taken regarding the question of whether protective policies are attractive or not. One view would be to suggest that the EU moves ahead with uniform carbon prices in all sectors of the region of concern, aiming primarily at meeting the domestic carbon target at the lowest possible cost, and hoping that such leadership inspires followers in the rest of the world and creates incentives for the development of more advanced technologies that can be exported.

The second view would be to argue in favour of the introduction of some form of protective policy so as to protect jobs or capital owners, or both. This article has reviewed some policies that aim at achieving these goals. Buying acceptance for climate policies might be important and necessary, but the policies used to protect the industries may be costly and they may also be difficult to get rid of (just witness the history of the common agricultural policy, introduced after World War II to ensure food production in Europe, but still in place). It is beyond the scope of this article to propose any solution to this trade-off, but it seems clear that more research is needed to develop policies that combine the conflicting objectives of being cost-efficient and politically feasible.

Acknowledgements

I am grateful to two anonymous reviewers, and to Dean Abrahamson, Robert Ayres, Göran Berndes, Fredrik Hedenus, Michael Hoel, Mike Hulme, Tomas Kåberger, Daniel Johansson, Per Kågeson, Malte Meinshausen, Bert Metz, Dennis Pamlin, Martin Persson, Karl-Henrik Robèrt, Thomas Sterner, Lars Zetterberg and Kerstin Åstrand, and to participants at the EFIEA Workshop on EU Strategy on Post-2012 Climate Policies held in Schevningen, The Netherlands, on 30–31 August 2004, and to members of the environmental advisory council of the Swedish government for inspiring discussions and/or critical comments on earlier versions of the manuscript. Thanks also to the Swedish Energy Agency and Formas for financial support.

Notes

1 Clearly, there is uncertainty about the long-run contribution from these gases but our assumption can be compared to the median value for the total value of the contribution from all non-CO_2 gases (including aerosols) in the SRES scenarios, which as estimated by Wigley (2004) is 1.5 W/m^2 (the 90% confidence interval is ±1 W/m^2). The SRES scenarios are base-case scenarios without any policy-driven reductions in the emissions of greenhouse gases (in order to mitigate climate change). With mitigation it is reasonable to assume that it is possible to get down to 1 W/m^2.

2 The study by Stainforth et al. (2005) reports a range of 2–11°C per CO_2-equivalent doubling, but there are rather compelling reasons to be cautious when interpreting the higher range. For instance, evidence related to changes in greenhouse gases during the last glacial era and the estimated temperature change suggests that it is unlikely that the climate sensitivity can be so high.

3 A region's share, $x_i(t)$, of the allowable global emissions is given by $x_i(t) = (1-t/50)\ E_i(2000)/E_{tot}(2000)+t/50\ P_i(2050)/P_{tot}(2050)$, where t is years after the year 2000, E and P are emissions and population in region i or in total.

4 Emissions of fossil carbon per capita in India, Indonesia and sub-Saharan Africa are 0.3, 0.3 and 0.1 tC/cap/year, respectively. Emissions of greenhouse gases including fossil carbon, methane and nitrous oxide calculated using 100 GWPs, are estimated at 0.5, 0.7 and 0.5 tC/cap/year, respectively; see Höhne (2005) based on UNFCCC.

5 Such differences are built into the Kyoto framework – the rich countries will have to take the lead – but similar problems can be expected for decades ahead since different countries view climate change differently and the alternative – to wait until everybody agrees that something should be done – would probably imply a rather long period of waiting. One approach could be to include all countries in a cap-and-trade system and distribute permits generously to those who resist so that they may end up being winners of the climate policy. This is roughly what happened with Russia in the Kyoto negotiations but it has so far – for both good and bad reasons – not received sufficient support to bring other countries on board in this manner.

6 The mirror image of this argument goes under the name the 'Dutch disease'; i.e. the fact that countries which experience an export boom in one sector (e.g. as a result of a discovery of petroleum) will see more resources drawn to that sector. The increase in export leads to an upward pressure on the exchange rate and to higher salaries in this sector, which leads to loss of competitiveness in other sectors.

7 Most of the carbon in the crude oil remains in the product, but there are some emissions in the refineries that, if taxed, would increase production costs and might lead to relocation of the refinery. It is in this context also worth observing a related problem: if ethanol, methanol, DME or FT diesel from biomass were to become competitive in Europe because of its carbon policies or the biofuels directive, it would be important to make sure that other regions did not produce the same fuels from fossil fuels (since that would be considerably cheaper, in particular for methanol and DME) and sell it as if it were fossil-carbon-free. Although the chemical composition of the fuels is the same regardless of the energy source, the isotopic content is different.

8 This is the case for large markets or islands (e.g. Australia, the EU, North America, and Iceland). But competition could also occur if, say, Turkey and Ukraine were to start selling large amounts of electricity to the EU as a result of climate policies in the EU, then countervailing measures would also have to be considered.

9 Bergman (1996) is an exception who finds that differentiated taxes will lead to lower costs to meet a domestic carbon target. He even concludes that 'differentiated taxes seem to be an almost perfect substitute to internationally coordinated taxes'.

10 They are also incapable of capturing non-equilibrium effects on energy markets, e.g. opportunities to increase energy efficiency and thus reduce carbon emissions at no costs (see, e.g., Ayres, 1994).

11 Furthermore, costs are almost exclusively measured in monetary terms, but the social costs of high unemployment rates in certain regions may also need specific attention.

12 Recently, they urged EU governments to block windfall profits from the EU Emissions Trading Scheme (see http://www.pointcarbon.com/article.php?articleID=4212&categoryID=279, Aug 8, 2004).

13 A coal-fired power plant is estimated to emit 225 gC/kWh (40% efficiency). At US\$100/tCO_2, this amounts to 0.8 cents/kWh. At a capacity factor of 75% the power plant would produce 6,570 kWh/year per installed kW. Thus, if the price increase can be passed on to consumers and the plant owner gets permits that correspond to its emissions, then the additional revenue is US\$53/kW of capacity/year. Assuming that the permit price increases with the discount rate, then 25 years of permits would be equal to US\$1,330/kW of installed capacity; more than the cost of building a coal-fired power plant!

14 This view is only partly valid. Policies to improve energy efficiency in cars or buildings, for instance, would not lead to lower emissions in a trading scheme in the current phase – that is certainly true – but it would lower the permit price, which in turn would make it possible for policy makers to adopt more stringent targets in subsequent periods.

15 This would not only lower the cost of meeting the climate target, but also bring about other environmental benefits associated with the reduction of mining and metals refining (see, e.g., Kåberger et al., 1994).

References

Alcamo, J., Kreileman, E., 1996. Emission scenarios and global climate protection. Global Environmental Change 6, 305–334.

Andronova, N.G., Schlesinger, M.E., 2001. Objective estimation of the probability density function for climate sensitivity. Journal of Geophysical Research 106, 22605–22611.

Arnell, N.W., Cannell, M.G.R., Hulme, M., Kovats, R.S., Mitchell, J.F.B., Nicholls, R.J., Parry, M.L., Livermore, M.T.J., White, A., 2002. The consequences of CO_2 stabilisation for the impacts of climate change. Climatic Change 53, 413–446.

Ayres, R.U., 1994. On economic disequilibrium and free lunch. Environmental and Resource Economics 4, 434–454.

Azar, C., Rodhe, H., 1997. Targets for stabilization of atmospheric CO_2. Science 276, 1818–1819.

Azar, C., Schneider, S.H., 2002. Are the economic costs of stabilizing the atmosphere prohibitive? Ecological Economics 42, 73–80.

Azar, C., Lindgren, K., Andersson, B., 2003. Global energy scenarios meeting stringent CO_2 constraints: cost effective fuel choices in the transportation sector. Energy Policy 31, 961–976.

Azar, C., Lindgren, K., Larson, E., Möllersten, K., 2005. Carbon capture and storage from fossil fuels and biomass: costs and potential role in stabilizing the atmosphere. Climatic Change (in press).

Babiker, M.H., Criqui, P. Ellerman, A.D., Reilly, J.M., Viguier, L., 2003. Assessing the impact of carbon tax differentiation in the European Union. Environmental Modelling and Assessment 8, 187–197.

Babiker, M.H., Bautista, M.E., Jacoby, H.D., Reilly, J., 2000. Effects of Differentiating Climate Policy by Sector: A United States Example. MIT Joint Program on the Science and Policy of Global Change, Report No 61. MIT, Cambridge, MA.

Baer, P., 2004. Probabilistic analysis of climate stabilization targets and the implications for precautionary policy. Paper presented at the American Geophysical Union Annual Meeting, 17 December 2004, San Francisco, CA.

Bergman, L., 1996. Sectoral differentiation as a substitute for international coordination of carbon taxes: a case study of Sweden. In: Braden, J.B., Folmer, H., Ulen, T.S., (Eds), Environmental Policy with Political and Economic Integration. Österreichische Energieagentur [Austrian Energy Agency], pp. 329–349.

Böhringer, C., Rutherford, T.F., 1997. Carbon taxes with exemptions in an open economy: a general equilibrium analysis of the German tax initiative. Journal of Environmental Economics and Management 32, 189–203.

Bollen, J., Manders, T., Veenendaal, P., 2004. How much does a 30% emission reduction cost? Macroeconomic effects of post-Kyoto climate policy in 2020. CPB Document No. 64. Netherlands Bureau for Economic Policy Analysis, The Hague, The Netherlands.

Bye, B., Nyborg, K., 2003. Are differentiated carbon taxes inefficient? A general equilibrium analysis. Energy Journal 24(2), 85–112.

Caldeira, K., Jain, A.K., Hoffert, M.I., 2003. Climate sensitivity uncertainty and the need for energy without CO_2 emission. Science 299, 2052–2054.

Carbon Trust, 2004. The European Emissions Trading Scheme: implications for industrial competitiveness. Carbon Trust [available at http://www.thecarbontrust.co.uk/carbontrust/].

Cole, M.A., 2004. Trade, the pollution haven hypothesis and the environmental Kuznets curve: examining the linkages. Ecological Economics 48, 71–81.

den Elzen, M.G.J., 2002. Exploring climate regimes for differentiation of future commitments to stabilize greenhouse gas concentrations. Integrated Assessment 3, 343–359.

den Elzen, M.G.J., Berk, M.M., 2004. Bottom-up Approaches for Defining Future Climate Mitigation Commitments. RIVM Report 728001029/2004. RIVM, Bilthoven, The Netherlands.

den Elzen, M., Lucas, P., van Vuuren, D., 2005. Abatement costs of post-Kyoto climate regimes. Energy Policy 33, 2138–2151.

EU [European Union], 2005. Council of the European Union, Presidency conclusions, March 22–23 [available at http://ue.eu.int/ueDocs/cms_Data/docs/pressData/en/ec/84335.pdf].

FEA [Federal Environmental Agency], 2003. Subsidizing Germany's hard coal is economically and ecologically detrimental. Press release 14/2003, Federal Environmental Agency, Germany [available at http://www.umweltbundesamt.de/uba-info-presse-e/presse-informationen-e/pe05703.htm].

Forest, C.E., Stone, P.H., Sokolov, A., Allen, M.R., Webster, M.D., 2001. Quantifying uncertainties in climate system properties with the use of recent climate observations. Science 295, 113–117.

Goulder, L.H., 2005. Reconciling cost-effectiveness and political-feasibility considerations in U.S. climate policy. Paper presented at the Whole Earth Systems: Science, Technology and Policy, Stanford University, February 10–12.

Grassl, H., Kokott, J., Kulessa, M., Luther, J., Nuscheler, F., Sauerborn, R., Schellnhuber, H.-J., Schubert, R., Schulze, E.-D., 2003. Climate Protection Strategies for the 21st Century: Kyoto and Beyond. Report prepared by the German Advisory Council on Global Change (WBGU), Berlin, Germany.

Gregory, J.M., Stouffer, R.J., Raper, S.C.B., Stott, P.A., Rayner, N.A., 2002. An observationally based estimate of the climate sensitivity. Journal of Climate 15, 3117–3121.

Grubb, M.J., Hope, C., Fouquet, R., 2002. Climatic implications of the Kyoto Protocol: the contribution of international spillover. Climatic Change 54, 11–28.

Grubb, M., Azar, C., Persson, M., 2005. Allowance allocation in the European emissions trading system: a commentary. Climate Policy 5(1), 127–136.

Ha-Duong, M., Grubb, M., Hourcade, J.C., 1997. Influence of socioeconomic inertia and uncertainty on optimal CO_2-emission abatement. Nature 390, 270–273.

Hansen, J.E., 2005. A slippery slope: how much global warming constitutes dangerous anthropogenic interference. Climatic Change 68, 269–279.

Hare, W., 2003. Assessment of Knowledge on Impacts of Climate Change. Contribution to the Specification of Article 2 of the UNFCCC: Impacts on Ecosystems, Food Production, Water and Socio-economic Systems. Report prepared for the German Advisory Council on Global Change [available at http://www.wbgu.de/wbgu_sn2003_ex01.pdf].

Hoel, M., 1996. Should a carbon tax be differentiated across sectors. Journal of Public Economics 59, 17–32.

Höhne, N.E., 2005. What is next after the Kyoto Protocol: assessment of options for international climate policy post 2012. PhD thesis, University of Utrecht, The Netherlands.

Hulme, M., 2003. Abrupt climate change: can society cope? Philosophical Transactions of the Royal Society of London Series A: Mathematical Physical and Engineering Sciences 361(1810), 2001–2019.

ICCT [International Climate Change Taskforce], 2005. Meeting the Climate Challenge: Recommendations of the International Climate Change Taskforce. American Progress Institute [available at http://www.americanprogress.org].

IPCC [Intergovernmental Panel on Climatic Change], 1994. Radiative Forcing of Climate Change and an Evaluation of the IPCC IS92 Emissions Scenarios. Houghton, J.T., Meira Filho, L.G., Bruce, J., Hoesung Lee, Callander, B.A., Haites, E., Harris N., Maskell, K., (Eds). Cambridge University Press: Cambridge, UK.

IPCC [Intergovernmental Panel on Climatic Change], 1996a. Climate Change 1995: The Science of Climate Change. Contribution of Working Group I to the Second Assessment Report of the Intergovernmental Panel on Climate Change. Houghton, J.T., Meira Filho, L.G., Callander, B.A., Harris, N., Kattenberg, A., Maskell, K. (Eds). Cambridge University Press: Cambridge, UK.

IPCC [Intergovernmental Panal on Climate Change], 1996b. Impacts, Adaptation and Mitigation Options. IPCC Working Group II. Cambridge University Press, Cambridge, UK.

IPCC [Intergovernmental Panel on Climatic Change], 2001a. Climate Change 2001. Mitigation Contribution of Working Group III to the Second Assessment Report of the Intergovernmental Panel on Climate Change. Metz, B., Davidson, O., Swart, R., Pan, J. (Eds). Cambridge University Press, Cambridge, UK.

IPCC [Intergovernmental Panel on Climatic Change], 2001b. Climate Change 2001. Impacts Adaptation and Vulnerability: Contribution of Working Group II to the Second Assessment Report of the Intergovernmental Panel on Climate Change. McCarthy, J.J., Canziani, O.F, Leary, N.A., Dokken, D., White, K.S. (Eds). Cambridge University Press, Cambridge, UK.

Jaffe, A.B., Peterson, S.R., Portney, P.R., Stavins, R.N., 1995. Environmental Regulation and International Competitiveness: What Does the Evidence Tell Us? Journal of Economic Literature, 33(1): 132–163.

Kåberger, T., Holmberg, J., Wirsenius, S., 1994. An environmental tax-shift with indirect desirable effects. International Journal of Sustainable Development and World Ecology 1, 250–258.

Kågeson, P., 2000. Europe's response to climate change: two scenarios. Working Paper. The Continue Project, SNS, Stockholm, Sweden.

Kerr, R.A., 2004. Three degrees of consensus. Science 305, 932–934.

Krugman, P., 1994. Competitiveness: a dangerous obsession. Foreign Affairs 73(2) [available at http://www.foreignaffairs.org/19940301faessay5094/paul-krugman/competitiveness-a-dangerous-obsession.html].

Lackner, K.S., 2003. A guide to CO2 sequestration. Science 300, 1677–1678.

Marland, G., Boden, T.A., Andres, R.J., 2003. Global, regional, and national CO_2 emissions. In: Trends: A Compendium of Data on Global Change. Carbon Dioxide Information Analysis Center, Oak Ridge National Laboratory, US Department of Energy, Oak Ridge, TN, USA [available at http://cdiac.esd.ornl.gov/trends/emis/tre_glob.htm].

Mastrandrea, M.D., Schneider, S.H., 2004. Probabalistic Integrated Assessment of 'Dangerous' Climate Change. Science 304: 571–575.

Meinshausen, M., 2005. On the risk of overshooting 2°C. Paper presented at the symposium Avoiding Dangerous Climate Change, Exeter, February 1–3 [available at http://www.stabilisation2005.com/programme.html].

Meyer, A., 2000. Contraction and convergence: the global solution to climate change. Schumacher Briefings 5. Greenbooks, Bristol, UK.

Nakicenovic, N., Riahi, K., 2003. Model runs with MESSAGE in the context of the further development of the Kyoto Protocol. Prepared for the report by Grassl et al. (2003).

National Board of Trade, 2004. Climate and Trade Rules: Harmony or Conflict? National Board of Trade, Stockholm [available at http://www.kommers.se/page_disp.asp?node=5].

Obersteiner, M., Azar, C., Kauppi, P., Möllersten, K., Moreira, J., Nilsson, S., Read, P., Riahi, K., Schlamadinger, B., Yamagata, Y., Yan, J., van Ypersele, J-P., 2001. Managing climate risks. Science 294, 786–787.

O'Neill, B.C., Oppenheimer, M., 2002. Dangerous climate impacts and the Kyoto Protocol. Science 96, 1971–1972.

Oppenheimer, M., Alley, R.B., 2004. The West Antarctic ice sheet and long term climate policy. Climatic Change 64, 1–10.

Oppenheimer, M., Alley, R.B., 2005. Ice sheets, global warming, and article 2 of the UNFCCC: an editorial essay. Climatic Change 68, 257–267.

Persson, M., 2003. Industrial migration in the chemical sector: do countries with lax environmental regulations specialize in polluting industries? Paper presented at the EAERE conference, Bilbao, June.

Persson, T.A., Azar, C., Lindgren, K., 2005. Allocation of CO_2 emission permits: economic incentives for emission reductions in developing countries. Energy Policy (in press) [available online from http://dx.doi.org/ at doi:10.1016/j.enpol.2005.02.001].

Porter, M.E., Van der Linde, C., 1995. Toward a New Conception of the Environment-Competitiveness Relationship. Journal of Economic Perspectives 9(4): 97–118.

Reinaud, J., 2004. Industrial competitiveness under the European Union Emissions Trading Scheme. IEA Information Paper. International Energy Agency, Paris.

Rial, J.A., Pielke, R.A., Sr., Beniston, M., Claussen, M., Canadell, J., Cox, P., Held, H., de Noblet-Ducoudré, N., Prinn, R., Reynolds, J.F., Salas, J.D., 2004. Nonlinearities, feedbacks and critical thresholds within the Earth's climate system. Climatic Change 65, 11–38.

Rijsberman, F.R., Swart, R.J. (Eds), 1990. Targets and Indicators of Climatic Change. Stockholm Environment Institute, Stockholm, Sweden.

Sandén, B., Azar, C., 2005. Near term technology policies for long term climate targets. Energy Policy 33, 1557–1576.

Schneider, S.H., Azar, C., 2001. Are uncertainties in climate and energy systems a justification for stronger near-term mitigation policies? Paper prepared for a Pew Center meeting on Timing of Climate Policies, Washington, October 11–12, 2001 [available at http://www.pewclimate.org/docUploads/timing%5Fazar%5Fschneider%2Epdf].

Schneider, S.H., Lane, J., 2005. An overview of dangerous climate change. Paper presented at the symposium Avoiding Dangerous Climate Change, Exeter, February 1–3 [available at http://www.stabilisation2005.com/programme.html].

Schneider, S.H., Kuntz-Duriseti, K., Azar, C., 2000. Costing nonlinearities, surprises and irreversible events. Pacific and Asian Journal of Energy 10(1), 81–106.

Stainforth, D.A., Aina, T., Christensen, C., Collins, M., Faull, N., Frame, D. J., et al., 2005. Uncertainty in predictions of the climate response to rising levels of greenhouse gases. Nature 433, 403–406.

WBGU [German Advisory Council on Global Change], 1995. Scenarios for Derivation of Global CO_2 Reduction Targets and Implementation Strategies. WBGU, Bremerhaven, Germany.

Wigley, T.M.L., 2004. Choosing a stabilization target for CO_2. Climatic Change 67, 1–11.

Wigley, T.M.L., Raper, S.C.B., 2001. Interpretations of high projections for global-mean warming. Science 293, 451–454.

Wigley, T.M.L., Richels, R., Edmonds, J., 1996. Economics and environmental choices in the stabilization of atmospheric CO_2 concentrations. Nature 379, 240–243.

UN, 1992. United Nations Framework Convention on Climate Change [available at http://unfccc.int/2860.php].

UN, 2004. World Population Prospects: The 2005 Revision Population Database [available at: http://esa.un.org/unpp/].

Climate Policy 5 (2005) 329–348

Post-2012 climate action in the broad framework of sustainable development policies: the role of the EU

Petra Tschakert[1]*, Lennart Olsson[2]

[1] Department of Geography/AESEDA, Pennsylvania State University, USA
[2] Lund University Centre for Sustainability Studies (LUCSUS), Lund University, Sweden

Received 27 January 2005; received in revised form 11 March 2005; accepted 11 March 2005

Abstract

The linkages between climate change and sustainable development are multiple and profound. Nonetheless, their respective policy regimes have so far evolved along parallel, if not competing, paths. What is lacking to date is a detailed conceptual understanding of the practicability of their integration through cross-sectoral policies and programmes. We propose a synergistic adaptive capacity (SAC) framework that places adaptive capacity and equity at the centre of current policy debates. This framework, based on social vulnerability as a linking element between climate change adaptation and poverty reduction, goes beyond current attempts to 'mainstream' adaptation and mitigation into national development priorities. We outline guidelines on how to operationalize the SAC framework and, at the end, define the role of the EU in promoting and implementing these synergies within the post-2012 climate policy regime.

Keywords: Climate change; Sustainable development; EU post-2012 policies; Synergistic adaptive capacity; Social vulnerability

1. Introduction

It is now starkly evident that the developing countries have a crucial role to play in responding to climate change. Emissions of greenhouse gases from developing countries are expected to rise exponentially in the coming decades. Moreover, the developing countries are, in many cases, likely to suffer most from the impacts of global climate change because of their limited resources to adapt, and thus their greater vulnerability. At the same time, climate change impacts are projected to be most severe in the tropics, where most of the vulnerable countries are located.

Responses to the threat of climate change are mitigation, adaptation, or a combination of both. Mitigation includes all measures aimed at reducing the risk of a dangerous interference with the climate system (e.g. emission reduction or CO_2 sequestration), while adaptation includes actions aimed at reducing the potential impact on society and nature from the consequences of climate change. One important difference between the two responses is that money spent on mitigation is

---

* Corresponding author. Tel.: +1-814-863-9399; fax: +1-814-863-7943
E-mail address: petra@psu.edu

a global public good with few tangible effects here and now, while money spent on adaptation is a local restricted good only, but with important tangible local effects for specific beneficiary groups. It can be argued that it is more difficult to promote mitigation actions rather than adaptation actions, although the latter are likely to intervene in the lives of millions of people and require careful multilayered assessments. Identifying different synergies, for example where adaptation also results in mitigation and vice versa, might be an effective way of promoting new responses.

Climate change and persistent poverty, the latter here understood as the flipside of sustainable development, can be described in the form of DPSIR schemes, which allow us to characterize the different responses taken by the current policy regimes. DPSIR is a theoretical framework for analysing environmental or other problems, based on five different types of characteristics (see Figure 1) where D stands for driving forces, P for pressure, S for state, I for impact, and R for responses (Smeets and Weterings, 1999). Effective policy regimes should ideally include all kinds of responses, from immediate adaptive measures to long-term actions directed towards the driving forces. The current climate policy regime, however, contains predominantly responses to 'pressure' (emission reduction) and to some extent to 'state' (carbon sinks). It lacks responses directed towards the driving forces and the impacts (adaptation). With respect to poverty reduction, the main emphasis has so far been on emergency relief, neglecting longer-term capacity building, adaptive learning, and state-level policy reforms.

A number of substantial funds and mechanisms related to the climate change and sustainable development policy regimes are now available, such as the Global Environment Facility (GEF), the Clean Development Mechanism (CDM), the Special Adaptation Fund for Least Developed Countries, different World Bank initiatives (Community Carbon Fund and Bio-Carbon Fund), and various forms of Official Development Assistance (ODA). Nevertheless, serious bottlenecks for creative action exist that restrain adequate policy responses in practice, including knowledge on appropriate and achievable activities and institutional capacity for implementation. What is lacking, most of all, is a clear notion of synergies.

The linkages between climate change and sustainable development are multiple and profound, even though their respective debates have, for a long time, evolved along rather separate paths. As stated by Najam et al. (2003b), it is no longer necessary to argue about the desirability of integrating climate change and sustainable development, due to the amply demonstrated evidence of their complementary and even synergistic relationship. Presumably, sustainable development has ceased being a simple 'add-on' to climate change policies or an 'in-the-bag' side-benefit to mitigation or adaptation activities. Rather, sustainable development should be seen as the driving force in the framing of the international climate debate.

In order to fully incorporate sustainable development into EU policy on climate change and assign it a leading role, clearly more specific emphasis is needed on integrated cross-sectoral policies and programmes. The three most urgent areas of EU policy on climate change and sustainable development are: (1) to promote adaptation for vulnerable groups and areas, (2) to underline clear and equitable connections between poverty eradication and climate policies, and (3) to counteract unsustainable patterns of consumption and production by promoting clean technology transfer to rapidly growing regions.

The aim of this article is to analyse the current and potential linkages between the climate change policy regime and the international regime on sustainable development, to outline the

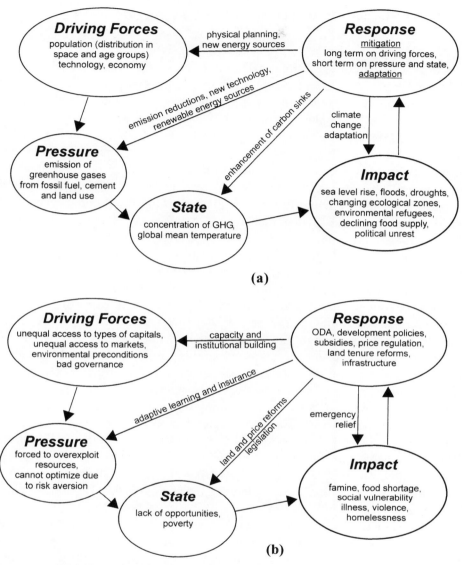

Figure 1. Climate change (a) and persistent poverty (b) expressed in the form of DPSIR schemes.

practicability of this integration embedded in a political economy framework, and to describe the role of the EU in operationalizing this linkage. We first provide a brief discussion of the major building blocks relative to the climate change–sustainable development nexus. Then we present concrete examples from forestry, energy, and transportation experiences within the CDM. In a third step, we propose a conceptual framework placing adaptive capacity and equity at the centre of the two policy regimes ('synergistic adaptive capacity'). Finally, based on this conceptual framework, we outline the possible role and position of the EU for the post-2012 period.

2. Building blocks of the climate change–sustainable development nexus

2.1. The UNFCCC and the IPCC

The UN Framework Convention on Climate Change (UNFCCC), signed in 1992 during the Earth Summit in Rio de Janeiro and entered into force in 1994, constitutes the foundation of global efforts to respond to global climate change. Its ultimate objective is the 'stabilization of greenhouse gas concentrations in the atmosphere at a level that would prevent dangerous anthropogenic interference with the climate system'. It is also in the UNFCCC that the policy mandate to tackle climate change within the context of sustainable development was first articulated, both as a right and an obligation (Najam et al., 2003b). In addition to the guiding precautionary principle that underpins the convention, 'common but differential responsibilities' assign the lead in combating climate change to the developed countries, while acknowledging the particular needs of developing countries, primarily with respect to sustainable and equitable development.

The apparent dilemma that results from these broadly defined principles is threefold. First, what is 'equitable' in the context of the 'common but differential responsibility' principle is resolved neither at the level of the North–South bargain (Najam et al., 2003a) nor between developing countries. Second, policies are constructed on the basis of 'rich' and 'poor' nations, but many 'poor' countries have increasing numbers of wealthy people with living standards and levels of consumption equal to that of many 'rich' countries. Third, 'dangerous', as stated in Article 2 of the convention, is a relative concept that might apply to one group but not to another. While only minor changes in the climate system may have disastrous impacts for some vulnerable groups or ecosystems, others might not perceive them as dangerous or, on the contrary, may see them as beneficial. Since the less-vulnerable groups are largely those in charge of mitigation responses, immediate action might not seem necessary. While there is still no agreement between developing and industrialized nations on what exactly constitutes 'dangerous' interference and how to efficiently and equitably operationalize the concept, there is no way around substantial cuts in global GHG emissions. So far, the Kyoto Protocol (KP), adopted in 1997 and entered into force on 16 February 2005, is the only international agreement in place that includes legally binding emission targets (Downing et al., 2003).

The Intergovernmental Panel on Climate Change (IPCC), the scientific body with a mandate to assess the state of knowledge on climate change and its impacts and possible response strategies, has been playing a major role in shaping the climate change policy debate. Nevertheless, it has so far discouraged an efficient integration of climate change into other major global social, economic, and financial concerns and sustainable development priorities. First, mitigation and adaptation policies have been largely considered to be unrelated and pursued along separate paths, and this is still evident in the separate reporting guidelines for national communications (Dang et al., 2003). Second, there is a seeming dichotomy between mainly biophysical impacts and mitigation strategies on one side, and socio-economic and institutional determinants of vulnerability and adaptation responses on the other side of the IPCC discourse. Especially during the first 'generation' of the IPCC assessment reports, clear emphasis was put on the reduction and stabilization of GHG as the major solution to climate change. However, there is growing recognition that climate change poses a serious threat to development as a whole, resulting in increasing emphasis on adaptation as one key solution. As expressed by Burton et al. (2002, p.146), adaptation has evolved from 'the handmaiden to impacts research in the mitigation context' to an activity that is now perceived as essential within the broader framework of sustainable development. Third, the IPCC, through its three assessment reports, has included equity and sustainable development concerns rather slowly and is still trying to 'catch up' with both of them (Najam et al., 2003b; Swart et al., 2003).

2.2. Mitigation and adaptation in the CDM

Among the three mitigation mechanisms (Emission Trading, Joint Implementation, and the Clean Development Mechanism (CDM)) set in place by the KP, the latter is the only policy instrument to encourage the participation of developing countries in climate change efforts. With its explicit dual mandate to simultaneously contribute to global climate change mitigation and national development objectives, specified in Article 12.2, the intrinsic nature of the mechanism as a win–win solution has been widely acclaimed by both Annex I and non-Annex I countries (Matsuo, 2003). Nevertheless, actual development benefits remain controversial (Brown et al., 2004).

Two of the most important, although contested, achievements with major implications for sustainable development of recent Conventions of the Parties (COPs) were the inclusion of afforestation and reforestation (sinks) into the CDM and the adoption of its definitions and modalities, respectively. So far, the major CDM-eligible programmes were those with a focus on renewable energy and sustainable transport. For non-Annex I countries, this new eligibility for forest-based carbon sink projects as part of the land-use, land-use change and forest activities (LULUCF) appears highly promising, partly because of potential benefits from carbon sequestration in marginal environments (Lal and Bruce, 1999; Olsson and Ardö, 2002). Yet, it is argued that elements essential to sustainable development are not sufficiently taken into account under the mandatory project design documents for sink projects under the CDM. One important aspect of CDM projects is that it is up to the recipient country to determine to what extent a project contributes to sustainable development. The contentious points are (unequal) distribution of knowledge, decision-making power, and anticipated economic and institutional benefits among project participants, as well as a potential gender bias (Karlsson and Clancy, 2000).

On the adaptation side of the climate change debate, significant progress has been made both conceptually and operationally. Adaptation activities are now supported through several funds and are intended to be country-driven, cost-effective, and integrated into national sustainable development and poverty-reduction strategies. National Adaptation Programmes of Action (NAPAs) and institutional infrastructure for training and capacity building are funded through the Least Developed Countries (LDC) Fund. US$16.5 million had been made available by donors as of April 2004 (http://unfccc.int/2860.php). Adaptation has been selected as the top priority in the Special Climate Change Fund (SCCF), followed by technology transfer.

2.3. The role of the IMF, the World Bank and GEF

The IMF and the World Bank are two of the most influential and important players among the international sustainable development institutions. In the context of climate change, however, their performance is controversial, despite new funding mechanisms. The IMF has recognized climate change as a major challenge for the future, including both mitigation and adaptation. But the analysis is simplistic and overly optimistic concerning the potential for technological solutions to adaptation in developing countries: 'technologies already exist or could be developed to address many of the adverse social and economic effects' (Heller and Mani, 2002, p.30).

The World Bank has been actively pursuing climate change in the context of development. Its former president, James D. Wolfensohn, expressed a compelling statement in this direction (Wolfensohn, 2004), although the Bank's course after the end of his second term remains a matter of debate: 'as one who has the privilege of representing an institution where we take this issue [climate change] very, very seriously, and where I'm delighted to have the chance to remind you that the issue is related to development'.

Nevertheless, this statement contrasts starkly with the Bank's actual funding strategy (Figure 2). In its portfolio of energy-related lending, fossil fuel projects comprised 98%, while renewable and energy efficiency projects accounted for only 2% during the 1995–2003 period. The World Bank often contributes very small funds to individual projects (e.g. in the controversial Chad–Cameroon oil pipeline project, the World Bank share is less than 5% of the total project) and thereby acts as a 'door-opener' to other funding organizations. While the overall contributions to energy projects have decreased, renewable energy and energy efficiency projects remain underrepresented in the Bank's lending portfolio. In terms of climate change adaptation, the World Bank has clearly expanded its portfolio (Mathur et al., 2004); the underlying assumption, however, seems to mirror the IMF viewpoint that adaptation is primarily a question of technological solutions and appropriate funding.

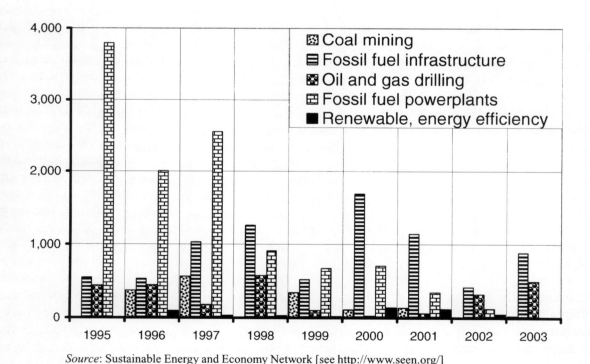

Source: Sustainable Energy and Economy Network [see http://www.seen.org/]

Figure 2. Lending from the World Bank Group, 1995–2003, for energy projects. The left-hand panel shows the overall distribution of lending into the five categories and the right-hand panel shows the lending profile over time.

The main purpose of the Global Environmental Facility (GEF) is to assist developing countries in meeting the objectives of international environmental conventions, mainly the UNFCC, CBD, UNCCD, and to promote sustainable livelihoods. The GEF has allocated US$4 billion in grants and leveraged an additional US$12 billion in co-financing from other sources to support more than 1,000 projects in over 140 developing nations and countries with economies in transition (http://www.gefweb.org/). Climate change is one of six GEF focal areas. However, among the 15 operational programmes for GEF grants, only one exists to date that encompasses cross-sectoral programmes (Integrated Ecosystem Management, OP12). So far, the experiences in linking GEF projects across environmental focal areas have resulted in only 'indirect' contributions to the enhancement of adaptive capacity. A more explicit focus on vulnerable groups and sectors and their differential adaptive capacity has been put forward recently through the GEF's strategic priority 'Piloting an Operational Approach to Adaptation' (SPA), stressing 'mainstreaming' of adaptation into sustainable development.

2.4. The sustainable development policy framework

On the side of the sustainable development policy regime, the major existing building blocks are Agenda 21, the Millennium Declaration, and the Johannesburg Plan of Action. These can be considered as a continuation of key principles put forward during the 1992 United Nations Conference on Environment and Development (UNCED). One of the most striking differences between the climate change and sustainable development policy regimes is the fact that, unlike the UNFCCC and the KP, the sustainable development objectives are defined as rather broad guidelines for global commitment without clearly specified targets and modalities for implementation. The various parties are mainly 'encouraged' or 'urged' to achieve the identified goals and targets, while no efficient mechanisms exist to ensure compliance.

The Millennium Declaration, adopted in 2000, is based on eight ambitious goals and numerous targets, among which the eradication of extreme poverty and hunger constitutes the top priority. Coping with climate change and managing natural resources in an efficient and equitable way have been identified as key areas essential for sustaining our global future. These targets and key areas clearly overlap with the recent adaptation emphasis in the climate change debate (in respect to both climate and non-climate stress) within more integrated and holistic frameworks on ecosystem management and sustainable development. However, for policy development and decision-making, critical synergies between climate change, global environmental goals, and development are likely to remain nebulous and vague.

The same argument can be made for the Johannesburg Declaration, adopted during the World Summit on Sustainable Development in 2002. Its three overarching objectives and essential requirements for sustainable development are: (1) poverty eradication, (2) changing unsustainable patterns of consumption and production, and (3) protecting and managing the natural resource base for economic and social development. Adaptation remains a rather latent concept, although the focus under Objective 3 is on an integrated and multi-hazard approach to address vulnerability, risk and disaster management, prevention, mitigation, response and recovery as fundamental elements of a safer world. A more explicit emphasis could also be made on equity, one of the three pillars of the EU sustainable development strategy and now also proposed as one conceptual prerequisite for successful integration of climate change and sustainable development for the

Fourth Assessment Report of the IPCC (Najam et al., 2003b). In particular with respect to unsustainable consumption and production patterns, there is little discussion on the polarity between rich and poor, differences in lifestyle, and future development pathways, not only between North and South but also between and within the developing countries themselves (Downing et al., 2003). For policy, it will be important to take a much more differentiated approach on equitable consumption and production, keeping in mind that the new emphasis on adaptation, primarily in developing countries, should not distract from a continuing focus on mitigation by major current and future emitters, both in Annex I and non-Annex I countries (Najam et al., 2003b).

As argued by (Najam et al., 2003a), efforts to combat global climate change and the pursuit of sustainable development can be seen as two sides of the same coin. In order to facilitate twin implementation, based on the operationalization of the identified synergies, two concerns need to be taken into account: (1) national and international responsibilities, and (2) adequate terminology.

Substantial efforts from both the international community and the developing countries will be necessary to implement Agenda 21 and achieve the internationally agreed development goals. Developed countries are urged to achieve the UN target of 0.7% of GNP in ODA and increase the flow of financial resources through international agreements and funding mechanisms. Only a few EU Member States, however, have reached the 0.7% target. Several Member States have not yet accomplished the Monterrey short-term commitment of 0.33%. Even if the EU policy target is clear, how to use the ODA remains ambiguous. There are certain risks that the official ODA is inflated by being combined with other kinds of external activities, such as technical cooperation, administrative and interest charges, debt forgiveness, and emergency relief. In the report from December 2003 (COM (2003) 829), the policy is extremely vague on this point: 'there is consensus among Member States to move doing more to coordinate development cooperation policies and harmonize procedures to take further action to untie aid, and to make the necessary provisions to ensure Member State participation in the HIPC'.

Despite financial commitments or promises from the North, the main responsibility for the design and implementation of appropriate development policies and strategies falls upon developing countries. Poverty reduction strategy papers (PRSP), prepared by individual low-income countries following a 1999 IMF and World Bank approach, are supposed to describe the countries' macroeconomic, structural, and social policies and programmes to effectively promote growth and reduce poverty. Developing countries are also to mobilize domestic resources, attract private investment, and utilize aid effectively. Without a clear perception on how to create these 'magical' synergies between national sustainable development priorities and global environmental commitments (combating desertification, mitigating climate change, and conserving biodiversity) and how to operationalize key linkages between NAPAs, PRSPs, and other national assessments most efficiently, this task risks being quite overwhelming.

Moreover, there is a need for simplified, shared language. Both the Millennium Development Goals and the Johannesburg Plan of Action cite as their top priority the eradication of extreme poverty, which can be equated with improved human well-being, and increased access to food, income, resources, energy and other basic needs. Now, from the perspective of the climate change policy regime, this is perceived as the cornerstone of improved adaptation or the more localized and short-term responses to a multitude of stressors, including climate change. However, due mainly to a rapidly growing body of literature, with a bewildering array of terms and concepts, the linkages between reduced poverty and vulnerability on the one side, and increased adaptive capacity on the other side of the policy regimes, are obscured. As a consequence, policymakers are likely to have a hard time determining the most promising synergies.

2.5. The role of the EU in the climate change–sustainable development debate

The EU climate change policy is firmly based on the IPCC Third Assessment Report. There is a broad consensus that climate change is a very serious threat to social, environmental and economic values in the EU and beyond. The 2003 heatwave in France, with over 10,000 deaths and over €10 billion in agricultural losses, is often taken as a concrete example of future climate change impacts. The EU policy is based on the following two principles: (1) the employment of market mechanisms in order to achieve emission reductions at the lowest possible cost, and (2) the involvement of all sectors of the economy. The EU has taken steps towards defining the meaning of Article 2 of the Kyoto Protocol, by setting a target of 2°C increase as the highest tolerable level. The main policy measures are the European trading scheme (ETS, started in January 2005) and a number of directives promoting end-user energy efficiency and eco-design in order to foster innovation of climate-friendly technologies.

The EU sustainable development strategy (SDS) was adopted by the European Council in Gothenburg in June 2001. Among its four key priorities, limiting climate change and increasing the use of clean energy is ranked first (the other three are addressing threats to public health, managing natural resources more responsibly, and improving the transport system and land use). The EU debate on sustainable development can be divided into two interlinked components: 'internal' and 'external'. The former focuses mainly on how to integrate environmental issues into other EU policy areas (Cardiff Process) and how to make EU policy more sustainable (Gothenburg Declaration). The latter concerns the EU's role in the global aspects of sustainable development, summarized in its final document 'Towards a Global Partnership for Sustainable Development'.

Among the key themes addressed in the SDS were the following: (1) the commitment to implement existing goals and targets, (2) 'appropriate national consultative processes' to consult widely with all stakeholders in order to harmonize national sustainable development strategies in the Member States, (3) the earliest possible date for the Kyoto Protocol to enter into force, (4) sustainable modes of transport, (5) new chemicals policy, and (6) improved biodiversity monitoring. The major focus of the global component was on poverty, including key points such as an increase of the EU's ODA to 0.39% of GNP, the central role of sustainable development in WTO agreements, and a 'coalition of the willing' for renewable energy sources and more sustainable patterns of production and consumption. In order to avoid segmentation, the EU strategy incorporates environment, equity and economy as the three key pillars of policies. Emphasis was placed on ways to spearhead EU efforts and to raise sustainable development standards at the global level to reconcile economic growth, social cohesion, and environmental protection. Another major issue was the integration of the new Member States from central and eastern Europe (CEE) by investing in more modern and more eco-friendly infrastructures and thereby decreasing their greenhouse gas emissions. One of the key challenges will be to assure effective sustainability synergies between environmental, social and economic dimensions.

3. Lessons from the AIJ and CDM projects

Novel attempts to create synergies between mitigation and adaptation policies and to integrate climate change measures and policies more efficiently into sectoral decision-making and development planning ('mainstreaming') have considerably transformed the climate change–sustainable development debate. Also new is the introduction of the twin concepts of 'adaptive' and 'mitigative' capacity. The debate has shifted from seeing sustainable development as a simple add-on, or even competitor, to climate change to a 'development-first approach' (Davidson et al., 2003), a broader focus on development with climate mitigation and other environmental goals as desirable ancillary benefits. By fusing the two debates, such a conceptual shift may considerably influence future policy responses. What seems to be lacking, however, is a creative, conceptually palpable, and cost-effective approach to integrate sustainable development into the long-term global mitigation strategies and the short-term localized adaptive responses to both climate change and other environmental stressors. In the next sections, we review some experiences on integrated climate–sustainable development projects (Activities Implemented Jointly (AIJ) and CDM) and then use the lessons for post-2012 EU climate action guidelines.

The first AIJ projects to be undertaken between Annex I and non-Annex I countries were launched under the pilot phase at COP-1 in 1995. All projects were required to be compatible with and supportive of national environmental and development priorities and strategies. These AIJ and early CDM projects have been serving as a test ground and a valuable learning experience for CDM projects that are only fully implemented once the KP enters into force. Out of a total of 157 projects, 20 dealt with agriculture and forestry and 137 with energy. A total of 76 projects were with EU new Member States, while 40 were with Latin American countries, 15 in Asia, and 10 in Africa (http://unfccc.int/kyoto_mechanisms/aij/activities_implemented_jointly/items/2094.php).

CDM projects have the dual mandate of reducing GHG emissions and contributing to sustainable development. Since Kyoto, ambitious claims have been made about the development benefits of market-based policy instruments such as the CDM. However, detailed research has remained scarce and benefits have often been shown to be more hypothetical than real (Brown and Corbera, 2003b; Brown et al., 2004).

One reason for the rather disappointing development achievements observed so far in the CDM learning phase is the fact that the new carbon economy is embedded in a discourse of global managerialism, creating a distorted focus on planned and external policy interventions while ignoring local ecological and social realities (Adger et al., 2001, 2003; Brown and Corbera, 2003a). Participation, decision-making, responsibilities and responses of local stakeholders, including potential winners and losers, tend to be overlooked. In other words, equity as a key component of sustainable development seems to have fallen through the cracks during project design and implementation. Another reason for the failure of many CDM pilot projects is related to the fact that markets usually do not both meet societal objectives and allocate resources efficiently. As a consequence, these newly emerging carbon markets may not readily provide the 'development' element prescribed by the CDM (Brown et al., 2004). The underlying reason, we argue, is that markets are unable to provide solutions for people who do not or cannot participate in the market. If we define the poor as 'people not included in society' (Wolfensohn, 1997), market mechanisms may not be an efficient approach to fight poverty. Even if market mechanisms can reach some of the poor, there is an obvious risk that such mechanisms increase inequality between those included and those excluded.

Among the 137 AIJ projects in the energy sector, 45% deal with energy efficiency and 40% with renewable energy, and a small proportion with fuel switching and fugitive gas capture (http://unfccc.int/kyoto_mechanisms/aij/activities_implemented_jointly/items/2094.php). The transport sector accounts for roughly 25% of the total GHG emissions and an even larger share of the emissions from many developing countries, but for only 5% of the CDM projects (Earth Negotiations Bulletin 13(2), December 2003; http://www.iisd.ca/climate/cop9/enbots/). There are several options for reducing emissions in the transport sector, including increased vehicle efficiency, improved GHG intensity of used fuel, reduced transport activity, alternative modes of transport and increased vehicle capacity. Difficulties in establishing baselines and calculating leakage are mainly responsible for the few CDM projects in the transport sector.

To date, there is no comprehensive assessment of development benefits within CDM projects nor of the ways in which these benefits are distributed (Brown et al., 2004). The most critical of the encountered bottlenecks and shortcomings are summarized in Box 1 for forestry sink projects.

Box 1: Examples from forestry and carbon sink projects

Results from AIJ pilot projects and CDM feasibility studies have shown that the initially envisioned win–win situations are rather illusionary. The major lessons learned comprise the following:

- Successful local development through CDM forestry projects depends on local contexts, history, and social and political relations, primarily land tenure and access to land (Brown et al., 2004);
- Very few substantial improvements so far to participating communities in pilot projects, in terms of income, diversification of production, and other environmental or development issues;
- Trade-offs between carbon gained through sequestration and local economic profitability rather than full win–win solutions; in other words, only very few agro-ecosystems or forest environments combine high carbon offset potential with high profitability (Sanchez, 2000; Palm et al., 2004; Tschakert, 2004);
- Additional trade-offs with respect to local social development, access to resources, and other environmental benefits, including erosion control and biodiversity conservation (Brown et al., 2004);
- Often unequal distribution of costs and benefits, leading to winners and losers at the local scale (Brown et al., 2004; Tschakert, 2004);
- Better-endowed local actors (clear property rights, access to robust institutional networks) are more likely to participate in decision-making processes and benefit from project outcomes;
- Actors who lack key means, including primarily marginalized groups such as indigenous people, women, and poor households, tend to be left out, the so-called 'innovativeness-need paradox' (Rogers, 1995);
- Lack of sufficient human and financial resources, training, and flexible, robust institutional structures (Klooster and Masera, 2000; Nelson and De Jong, 2003; Brown et al., 2004);
- Shift from broad development emphasis to individual carbon focus (Nelson and de Jong, 2003);
- Concept of carbon markets is difficult to understand for local communities (Brown et al., 2004);
- By targeting mainly innovators and risk-takers (who can afford it) rather than vulnerable and risk-avert groups, CDM projects exacerbate rather than smooth out existing social inequalities.

The major lesson learned from the existing AIJ experiments is that it is simply not enough to count on sustainable development as a side-benefit or by-product of CDM mitigation projects. Some examples actually show that inequality has been reinforced rather than decreased. Given the current carbon market dynamics, positive records are even less likely in the future.

Based on the lessons from AIJ and CDM projects, we argue that an explicit focus on adaptation and sustainable development is needed in future climate change programmes, that equity should be taken into account, and that conflicts over decision-making, participation, and the distribution of benefits are addressed from the very beginning. This should include an efficient approach to rigorously assess trade-offs between the social, economic and environmental project criteria. This appears most feasible if undertaken through the lens of a social vulnerability framework that focuses explicitly on enhanced adaptive capacity and risk management.

4. A conceptual framework of synergistic adaptive capacity

In this section we take a closer look at adaptation as a key synergistic element between climate change and sustainable development. We propose a conceptual framework embedded in political economy thinking and with adaptive capacity and equity at its centre. It is a framework that the EU could use to better align its policy architecture for the post-2012 climate negotiations.

Although adaptation is now receiving increasing emphasis in climate change and development discourses, there is no need to 'reinvent' the concept as such or recreate it from scratch. Humans have always adapted to stress and shocks, more or less successfully. By grounding adaptation in the present (Burton et al. 2002), we observe what individual actors and institutions do and what the social, economic and environmental consequences of their actions are. In general, we can differentiate between spontaneous, planned and forced adaptation (see examples in Table 1). The latter refers to cases where a few 'lucky survivors' force upon a large mass of 'losers' a type of adaptation that is ethically unacceptable, including individual deaths. Policies and practices that are unrelated to climate but which still increase a system's vulnerability to climate change are termed 'maladaptation' (Burton, 1996, 1997).

The above examples of adaptation to climatic stressors may be socially, environmentally or economically desirable, but lack two key components when perceived through a sustainable development lens: (1) a clear understanding of the social and economic processes that facilitate and constrain adaptation, and (2) a differentiated perception on whether or not the outcomes are equitably distributed. In other words, processes of adaptation are intrinsically linked to the wider political economy of uneven development, resulting in winners and losers (Adger et al., 2003).

Hence, we propose to discuss adaptation not only within the context of biophysical stressors, impacts, and prevention (hazard events and outcomes) but also, or even primarily, as the flipside of what Allen (2003) refers to as 'structural, underlying vulnerability', a 'contextual weakness or susceptibility underpinning daily life'. It refers to the 'social vulnerability' which is system-specific and largely determined by poverty, inequality, and marginalization (Brooks, 2003). The advantages of such a *'social vulnerability approach'* for policy regimes are two-fold. First, it allows focusing more consciously on people as agents of change who are constantly managing risk and adapting to some kind of stress, based on their differential asset base and access to resources, thus drawing upon a set of experiences. This implies that successful adaptation does not necessarily have to be

Table 1. Different types of adaptation

Climatic stressor	Adaptive response	Type	Result (good vs bad)
Increased frequency of floods in a city	Purchase of SUV	Spontaneous, individual	Negative – increased emissions
Increased frequency of floods in a city	Management of vegetation and land use in river basin	Planned	Positive – carbon sequestration
Heat wave	Planting trees to shade houses	Spontaneous, individual	Positive – carbon sequestration
Heat wave	Purchase of AC	Spontaneous, individual	Negative – increased emissions
Drought	Out-migration	Spontaneous, individual	?
Drought	Emergency food	Planned (policies)	?
Hurricane	Timely evacuation of the well-endowed	Spontaneous, individual	Positive and negative
Hurricane, drought or flood	Inadequate response resulting in losses of life and property	Forced, maladaptation	Negative, but the 'lucky survivors' stand a better chance next time, due to reduced pressure on resources

planned at the top and be executed in a managerial way, but it is carried out as individual and spontaneous action that can be reinforced through the enhancement of people's resilience to cope with an uncertain future. The literature provides ample examples from soil and water conservation and agroforestry projects that have succeeded because people's capacity and innovativeness were supported rather than pre-made and 'one-size-fits-all' technical solutions (Mazzucato and Niemeijer, 2000; McDonald and Brown, 2000; Scoones, 2001; Franzel and Scherr, 2002).

Second, a 'pro-poor' and integrated approach, as promoted by UNDP, the World Bank, and other development institutions such as DFID, places poor and vulnerable people at the centre of the analysis, where they are seen as part of the solution rather than the problem for improving environmental management and reducing poverty. Such an approach attaches a more active connotation to the concept of poverty eradication, dropping the still prevailing perception of large regions of the world as 'disease-ridden, poverty-stricken, and disaster-prone', with inhabitants incapable of removing themselves from danger and destitution (Pelling, 2003). As argued by (Christoplos, 2003), poor people tend not to use their livelihood strategies to 'escape' from poverty *per se* but rather to cope with risks and shocks and address structural vulnerability.

The 'social vulnerability approach' also appears to be highly useful when addressing the inherent dilemma between basic survival strategies and more complex development options. The former often aim to simply reduce poverty and risk aversion in the short run, but deteriorate into dead-end solutions in the long run ('missionary hand mill' mode). In contrast, the latter imply more flexible and diverse opportunities to cope with risks and develop adaptive responses at a larger scale ('livelihood empowerment' mode). The evolution of emerging synergies as well as current shortcomings are summarized in Box 2.

Box 2: Emerging synergies between climate change and sustainable development based on a 'pro-poor' focus on vulnerability and adaptation

- Socio-economic development patterns determine the vulnerability to climate change and the human capacity for mitigation and adaptation (Klein et al., 2003);
- Present livelihood and risk management strategies are an efficient entry point for assessing and enhancing adaptive capacity and measures relating to climatic and non-climatic stressors;
- Social capital and strong institutions that address socio-economic and environmental problems often also improve people's capability to mitigate climate change or adapt to it (Swart et al., 2003);
- There is an apparent shift from activities to social actors and their capacity to simultaneously contribute to mitigation and adaptation (mitigative and adaptative capacity);
- The adoption of an 'ecosystem approach' (integrated management of land, water, and living resources) provides a more holistic understanding of ecosystems and their social actors.

Current shortcomings

- Synergies between mitigation options, adaptation benefits, and sustainable development are assessed mostly qualitatively, based on best guesses (low–high, positive–negative);
- Strong bias towards technical solutions rather than institutional strategies focusing on human agency;
- Focus on 'produced' adaptation, as an indirect contribution from integrated ecosystem management rather than an explicit policy priority that is oriented towards the enhancement of adaptive capacity;
- Some alarming parallels to sustainable development, unrealistically perceived as a by-product of CDM projects without a clear emphasis on equity.

We argue that synergies between climate change and sustainable development ought to be based on the 'livelihood empowerment' mode and linked to reduced social vulnerability. To make these synergies more apparent, we propose a new conceptual framework entitled '*synergistic adaptive capacity*' (SAC). This substitutes poverty eradication, as put forward in the general development policy regime, with increased adaptive capacity/reduced social vulnerability. As such, enhanced adaptive capacity constitutes not only an important response to climate change but is also a key to sustainable and equitable global environmental management. Figure 3 depicts this synergistic role (with adaptation in the middle). An equity filter ensures a more equitable distribution of roles and responsibilities on four levels: short-term adaptation, long-term mitigation, production and consumption, and environmental sustainability.

The SAC framework illustrates that adaptation and adaptive capacity are key to resilient livelihoods, reduced social vulnerability, and strengthened risk management. At the same time, improved adaptive capacity to a variety of stressors, especially among the currently most vulnerable

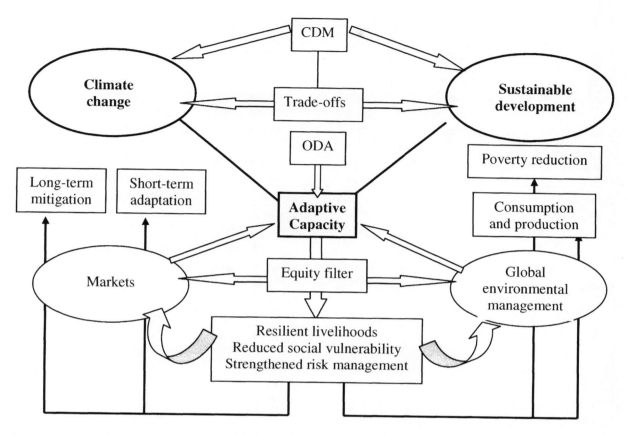

Figure 3. A conceptual illustration of the 'synergistic adaptive capacity' (SAC) framework.

rural and urban populations, is likely to contribute to better adaptation to climatic stressors and more efficient mitigation strategies. It is also expected to integrate the excluded poor more effectively into markets and contribute to more sustainable environmental management. Although current efforts to 'mainstream' mitigation and adaptation policies into development planning constitutes a step in the right direction, it seems to put the cart before the horse, following an overly simplistic efficiency thinking rather than a holistic conceptual outline.

A crucial element for promoting the synergistic adaptation paradigm, at the level of both policies and ODA, is to adopt a livelihood diversification approach and focus specifically on differential access to natural, human, productive, social, and institutional capital. Many of the CDM projects reviewed perceive poor communities as homogeneous entities, ignoring inequality at the lowest level, the household (Brown and Corbera, 2003a). A 'pro-poor' approach means deliberately working with those who have the least adaptive potential to promote strategies that could reverse this trend, even if overall economic or ecological gains are not maximized, at least in the short run. It also means drawing a line between contexts where adaptation is feasible and those where relief might be the only immediate option.

5. EU post-2012 policy outlook on integrated climate change and sustainable development

5.1. Broad integration and synergies

In this final step, we present guidelines to operationalize the SAC framework within the context of the EU's role in the post-2012 climate policy regime. We believe that SAC helps identifying effective synergies between environmental, social, and economic dimensions of both the climate and development policy regimes. It reframes the climate change debate from an environmental-economic issue to a more explicit development issue guided by the notion of sustainability and building on the EU's 'Towards a Global Partnership for Sustainable Development'. The Commission's Communication on Climate Change in the Context of Development Cooperation from March 2003 and the Action Plan 2004–2008 are indications pointing in this direction.

During the first commitment period of the KP, adaptation is barely taken into account. However, we argue that it should be a central element of the post-2012 policy regime and the EU should play a crucial role in ensuring that it happens. Discussions, then, may centre around crucial issues such as 'poverty-reduction targets' instead of 'emission reduction targets', and 'technical and institutional options' instead of 'burden-sharing'. Shifting the emphasis from impacts to solutions also implies a larger probability of getting developing countries committed to climate change discussions and actions. After all, development and poverty reduction, rather than emission reduction, are still the foremost priorities of the South (Biermann, 2005).

The EU should take a leading role in implementing the proposed SAC approach by initiating, designing, funding and monitoring programmes that implicitly follow a livelihood empowerment strategy, both in the new Member States and in developing countries. This implies addressing adaptive capacity, development, equity, and global environmental goals simultaneously, based on a 'pro-poor' and integrated cross-cutting approach focusing on both climate and non-climate stressors. There is an urgent need to develop explicit methods and indicators for assessing progress and measuring outputs on the ground as well as suitable funding schemes for adaptation that focus equally on sustainable livelihoods and sustainable ecosystems.

Moreover, for the EU it will be crucial to promote and operationalize synergies between three main international conventions (UNFCCC, Convention on Biological Diversity, and Convention to Combat Desertification), recognizing adaptation as a key link while emphasizing the efficacy of the ecosystem approach in addressing climate change and adaptive capacity. With or without the participation of the USA in a post-2012 climate policy regime, it can be argued that Europe will be sufficiently committed to further support policies and programmes that will steadily promote the implementation of the UNCCD. At the same time, the EU holds the potential to shape its climate policies in accordance with its current flagship strategies, deliberately pursuing fusion rather than segmentation. Institutional capacity building, adaptive and collaborative learning, and equitable legislative measures, as outlined as response strategies to persistent poverty (see Figure 1b), are also crucial in biodiversity conservation, desertification control, and climate change mitigation. Consequently, actions to combat desertification and to enhance the resilience of agro-ecosystems are a vital prerequisite for increased food security and improved well-being, not only in the developing world. Thus, the EU will have to design policies that transgress the conventional mainstreaming of the conventions' goals and instruments into national and regional development priorities. The challenge is, again, to design appropriate metrics that are transparent, holistic and accessible.

5.2. Crucial elements of a sustainable development-focused climate policy

In order to stress clear and equitable connections between poverty eradication, enhanced adaptive capacity, and climate policies, it will be vital for the EU to embed its policy regime within the larger political economy of development and over longer time frames. For this purpose, a clear vision of sustainable development is needed, a vision that is also shared by those who are the anticipated beneficiaries in the South. Such a vision is likely to focus on people's improved capacity to deal with risks, stresses, and shocks and to increase their underlying social resilience rather than escaping poverty *per se* or adapting exclusively to climatic changes, detached from their overriding livelihood concerns,. Hence, sustainable development cannot be a desirable endpoint, but only an ongoing process rooted in the present.

Moreover, to implement the SAC framework and to make it useful to a wide range of decision-makers and stakeholders in a participatory way, ODA needs to be harnessed to build local and regional institutional capacity. This includes some form of facilitated reaction of developing countries to shifting paradigms and synergistic policy regimes as well as training on to efficiently integrate national assessments, inventories, and action plans. Such a learning process, in fact, appears indispensable for both the climate change and the development community. So far, they seem to have opted for competition rather than collaboration.

The EU also has a major role to play in transforming experiences from the first CDM learning phase into more equitable project guidelines and activities, following trade-off analysis methods such as the one proposed by Brown et al. (2004), based on the weighting of the climate, social, economic and environmental criteria by direct and secondary stakeholders. For CDM projects to be effective in enhancing adaptive capacity, there is a strong necessity for supporting actions by which participation can be promoted on a fair and equal basis. These supporting actions cannot be market-based, but need to be funded outside the CDM, for instance through ODA. An incremental costs system, as currently used by GEF, could be implemented to enforce an integrated and equitable ecosystem approach on existing development initiatives and *vice versa*. Broad compliance with SAC can then be valued as incremental benefits, supported by funds set aside for this purpose, and applied to relevant EU programmes. Several of the AIJ and CDM projects reviewed in Box 1 could have been more successful in terms of poverty reduction and equity if they had been backed up by actions specifically aimed towards providing the necessary prerequisites for participation.

In addition, integrated climate change and development projects should include special mechanisms that exempt small-scale communities, marginal populations, and vulnerable groups from competing monitoring and verification services whose ethical stimuli are increasingly disputed. It is worth noting that, so far, such services are provided exclusively from institutions in the North. A first attempt is made by negotiating rules and regulations for small-scale projects, but the EU could play a much larger role in facilitating these efforts. Also, it would be highly desirable if the EU decided to support monitoring and verification services held by developing countries, either through the provision of infrastructure, institutional capacity building, or improved networking with European-based structures. Such measures would transform the often high transaction costs of CDM projects into local job opportunities, and thereby address some of the most common criticisms of CDM.

Finally, the EU to date is in an excellent position to counteract unsustainable patterns of consumption and production. Because of the current policy regime, built upon nation states and groups of nation states, there is very limited scope for addressing consumption in developing countries. The current trend of increasing income disparities in rapidly developing countries will most likely persist into the foreseeable future. This means that there will be very large groups of people in 'poor countries' with living standards and consumption patterns (including emission of GHG) on a par with North America and Europe. One approach to address this policy dilemma is through technology innovation. One can argue that the developer of a technology should be given a responsibility for how the specific technology is used and what consequences its use may have on third parties. From an EU point of view, very strict environmental regulations, for example on automobiles, would most probably be in the interests of the EU car industry in the long term. Car manufacturers in East Asia are becoming increasingly competitive due to substantially lower production costs. Very strict environmental regulations (e.g. zero-emitters) would probably favour the mature and technically very advanced European car industry.

6. Conclusions

In the EU, work on the post-2012 climate change policy is already under way. The most significant difference compared to the current policy regime is probably that adaptation will become equally as important as mitigation. There is a growing consensus that even with substantial and successful mitigation, we will not be able to avoid serious climate impacts in the coming decades. However, it would be a serious mistake, particularly with respect to developing countries, if this new emphasis on adaptation shifted the main focus away from mitigation. Instead, we must find new and innovative ways of continuing a strict emission reduction path and, at the same time, promote adaptation. One way of achieving this dual goal is to identify and promote realistic synergies of adaptation and mitigation. This might also be a successful way of more actively involving developing countries in a global agreement that effectively addresses the serious threat of climate change.

References

Adger, W.N., Benjaminsen, T.A., Brown, K., Svarstad, H., 2001. Advancing a political ecology of global environmental discourses. Development and Change 32, 681–715.

Adger, W.N., Huq, S., Brown, K., Conway, D., Hulme, M., 2003. Adaptation to climate change in the developing world. Progress in Development Studies 3(3), 179–195.

Allen, K., 2003. Vulnerability reduction and the community-based approach. In: Pelling, M. (Ed.), Natural Disasters and Development in a Globalizing World. Routledge, London.

Biermann, F., 2005. Between the USA and the South: strategic choices for European climate policy. Climate Policy 5, this issue.

Brooks, N., 2003. Vulnerability, risk and adaptation: a conceptual framework. Working Paper 38. Tyndall Centre for Climate Change Research.

Brown, K., Corbera, E., 2003a. Exploring equity and sustainable development in the new carbon economy. Climate Policy 3[Supplement 1], S41–S56.

Brown, K., Corbera, E., 2003b. A multi-criteria assessment framework for carbon-mitigation projects: putting 'development' in the centre of decision-making. Working Paper 29. Tyndall Centre for Climate Change Research.

Brown, K., Adger, W.N., Boyd, E., Corbera-Elizalde, E., Shackley, S., 2004. How do CDM projects contribute to sustainable development? Technical Report 16. Tyndall Centre for Climate Change Research.

Burton, I., 1996. The growth of adaptation capacity: practice and policy. In: Smith, J.B., Bhatti, N., Menzhulin, G.V., Benioff, R., Campos, M., Jallow, B., Rijsberman, F., Budyko, M.I., Dixon, R.K. (Eds), Adapting to Climate Change: An International Perspective. Springer-Verlag, New York, pp. 55–67.

Burton, I., 1997. Vulnerability and adaptive response in the context of climate and climate change. Climatic Change 36(1–2), 185–196.

Burton, I., Huq, S., Lim, B., Pilifosova, O., Schipper, E.L., 2002. From impacts assessment to adaptation priorities: the shaping of adaptation policy. Climate Policy 2, 145–159.

Christoplos, I., 2003. Actors at risk. In: Pelling, M. (Ed.), Natural Disasters and Development in a Globalizing World. Routledge, London, pp. 95–109.

Dang, H.H., Michaelowa, A., Tuan, D.D., 2003. Synergy of adaptation and mitigation strategies in the context of sustainable development: the case of Vietnam. Climate Policy 3[Supplement 1], S81–S96.

Davidson, O., Halsnaes, K., Huq, S., Kok, M., Metz, B., Sokona, Y., Verhagen, J., 2003. The development and climate nexus: the case of sub-Saharan Africa. Climate Policy 3[Supplement 1], S97–S113.

Downing, T.E., Munasinghe, M., Depledge, J., 2003. Special supplement on climate change and sustainable development. Climate Policy 3[Supplement 1], S3–S8.

Franzel, S., Scherr, S.J. (Eds), 2002. Trees on the farm: assessing the adoption potential of agroforestry practices in Africa. CABI Publishing, Wallingford, UK.

Heller, P. Mani, M., 2002. Adapting to climate change. Finance and Development 39(1), 29–31.

Karlsson, G., Clancy, J., 2000. Gender and energy: how is gender relevant to sustainable energy policies? In: Takada, M., Morris, E., Rajan, S.C. (Eds), Sustainable Energy Strategies: Materials for Decision-Makers. UNDP, New York.

Klein, R.J.T., Schipper, E.L., Dessai, S., 2003. Integrating mitigation and adaptation into climate and development policy: three research questions. Working Paper 40. Tyndall Centre for Climate Change Research.

Klooster, D., Masera, O., 2000. Community forest management in Mexico: carbon mitigation and biodiversity conservation through rural development. Global Environmental Change 10(4), 259–272.

Lal, R., Bruce, J.P., 1999. The potential of world cropland soils to sequester C and mitigate the greenhouse effect. Environmental Science and Policy 2(2), 177–185.

Mathur, A., Burton, I., van Aalst, M., 2004. An Adaptation Mosaic: A Sample of the Emerging World Bank Work in Climate Change Adaptation. World Bank Global Climate Change Team.

Matsuo, N., 2003. CDM in the Kyoto negotiations: how CDM has worked as a bridge between developed and developing worlds? Mitigation and Adaptation Strategies for Global Change 8, 191–200.

Mazzucato, V., Niemeijer, D., 2000. Rethinking Soil and Water Conservation in a Changing Society: A Case Study in Eastern Burkina Faso. Wageningen University, Wageningen, The Netherlands.

McDonald, M., Brown, K., 2000. Soil and water conservation projects and rural livelihoods: options for design and research to enhance adoption and adaptation. Land Degradation and Development 11, 343–361.

Najam, A., Huq, S., Sokona, Y., 2003a. Climate negotiations beyond Kyoto: developing countries' concerns and interests. Climate Policy 3, 221–231.

Najam, A., Rahman, A.A., Huq, S., Sokona, Y., 2003b. Integrating sustainable development into the Fourth Assessment Report of the Intergovernmental Panel on Climate Change. Climate Policy 3[Supplement 1], S9–S17.

Nelson, K.C., De Jong, B.H.J., 2003. Making global initiatives local realities: carbon mitigation projects in Chiapas, Mexico. Global Environmental Change 13, 19–30.

Olsson, L., Ardö, J., 2002. Soil carbon sequestration in degraded semiarid agro-ecosystems: perils and potential. Ambio 31, 471–477.

Palm, C., Tomich, T., Van Noordwijk, M., Vosti, S., Gockowski, J., Alegre, J., Verchot, L., 2004. Mitigating GHG emissions in the humid tropics: case studies from the alternatives to slash-and-burn program (ASB). Environment, Development and Sustainability 6, 145–162.

Pelling, M., 2003. Paradigms of risk. In: Pelling, M. (Ed.), Natural Disasters and Development in a Globalizing World. Routledge, London, pp. 3–16.

Rogers, E.M., 1995. Diffusion of Innovation. Free Press, New York.

Sanchez, P.A., 2000. Linking climate change research with food security and poverty reduction in the tropics. Agriculture, Ecosystems and Environment 82(1–3), 371–383.

Scoones, I. (Ed.), 2001. Dynamics and Diversity: Soil Fertility and Farming Livelihoods in Africa: Case Studies from Ethiopia, Mali, and Zimbabwe. Earthscan Publications, London.

Smeets, E., Weterings, R., 1999. Environmental Indicators: Typology and Overview. European Environment Agency, Copenhagen.

Swart, R., Robinson, J., Cohen, S., 2003. Climate change and sustainable development: expanding the options. Climate Policy 3[Supplement 1], S19–S40.

Tschakert, P., 2004. The costs of soil carbon sequestration: an economic analysis for small-scale farming systems in Senegal. Agricultural Systems 81, 227–253.

Wolfensohn, J.D., 1997. The challenge of Inclusion. Opening Address, World Bank Board of Governors Annual Meeting, Hong Kong SAR, China, 23 September, 1997.

Wolfensohn, J.D., 2004. Remarks on the Environment. Speech at the Brookings Institution, Washington, DC, 25 June 2004.

Climate Policy 5 (2005) 349–361

The European Union and future climate policy: Is mainstreaming adaptation a distraction or part of the solution?

Farhana Yamin*

Institute of Development Studies, University of Sussex, UK

Received 24 January 2005; received in revised form 10 March 2005; accepted 10 March 2005

Abstract

This article reviews the European Union's stance and policies on climate change adaptation and argues that developing a coherent long-term European strategy on climate change post-2012 will require the European Union to focus more strongly on adaptation issues than has hitherto been the case. It suggests that the EU should examine the dissonance between its prescriptions for integrating adaptation within the EU with its prescriptions to developing countries to mainstream adaptation. The EU should avoid a carrot-and-stick approach to adaptation funding and should focus on identifying common institutional and learning challenges with developing countries.

Keywords: Climate Policy; Adaptation; Vulnerability; Future commitments

1. Introduction

The 2001 Third Assessment Report of the Intergovernmental Panel on Climate Change (IPCC) confirms that the world's climate is changing and that some adverse impacts cannot be averted (IPCC, 2001). The IPCC has consistently pointed out that these impacts will fall disproportionately on developing countries and will adversely affect the poorest and most vulnerable groups the worst (Pachauri, 2004). These impacts are so significant that a recent assessment by the donor community has confirmed that climate change is a serious risk to poverty reduction and threatens to undo decades of development, potentially jeopardizing achievement of the Millennium Development Goals, in particular those related to the eradication of poverty and hunger, and to health and sustainable development (Sperling, 2003).

Adaptation has always been an integral component of the 1992 United Nations Framework Convention on Climate Change (UNFCCC). During the Convention's negotiations, for example, the Alliance of Small Island States (AOSIS) demanded legally-binding mitigation targets for industrialized countries to cut emissions of greenhouse gases (GHGs) as well as provisions mandating that richer countries provide assistance to vulnerable countries to help them meet their adaptation-related costs (Yamin, 2004). More recently, Least Developed Countries (LDCs) and

---

* Corresponding author. Tel.: +44-1273-877369
E-mail address: F.Yamin@ids.ac.uk

larger developing countries with sizeable climatically vulnerable populations, such as India, have also begun to focus on adaptation, with the result that the last three sessions of the UNFCCC Conference of the Parties (COP) have been termed 'adaptation COPs' (COP-10, 2004). As the recent floods and heatwaves in Europe demonstrate, however, richer countries cannot simply opt out of costly climate impacts (Linnerooth-Bayer and Amendola, 2003; World Disasters Report, 2004). Worryingly, adverse climate impacts appear to be on the increase in Europe, as pointed out in the European Environment Agency's 2004 Report, mirroring a broader trend worldwide (EEA, 2004a; Poumadere and Le Mer, 2004).

Despite the higher scientific and political profile given to adaptation in recent years, most policy discussions and dialogues relating to 'next steps' continue to focus disproportionately on the next round of mitigation commitments; in particular how climate policy can tackle growing US emissions in the face of policy inaction at the federal level and the willingness of developing countries to deepen implementation of mitigation policies.[1]

This article suggests that because adaptation will be a major concern for all developing countries for decades to come, it will be a core issue in the 'next steps' discussions that will commence in May 2005 with the convening of the seminar of governmental experts (SOGE) as agreed upon at the tenth session of the COP in Argentina. The article argues that the development of a coherent long-term European strategy on climate change post-2012 will require the EU, and other negotiating partners, to focus much more deeply on adaptation issues than has hitherto been the case. Advocating developing countries' mainstream climate adaptation will require, in turn, that the EU needs to pursue policy integration within the EU to ensure that environmental considerations, specifically climate protection and adaptation policies, are integrated into other policy areas – a process which is currently at an early stage in the EU (EEA, 2004b).

2. Adaptation and the international climate regime

Adaptation to the adverse effects of climate change has been recognized as an important element of the climate regime since its inception. Although industrialized countries initially resisted their inclusion, developing countries, led primarily by AOSIS on this issue, insisted that adaptation provisions be included in the Convention.[2] Accordingly, the Convention contains commitments relating to the prevention of the adverse effects of climate change as well as financial assistance from richer countries to support developing countries in meeting adaptation costs (Verheyen, 2002; Yamin and Depledge, 2004).

Implementation of the Convention's adaptation provisions has been impeded, however, by three interlocking factors:

- Lack of agreement about the meaning, scope and timing of adaptation
- Limited institutional capacity in developing countries to undertake vulnerability assessments and adaptation planning (V&A)
- Bottlenecks in the availability of funding for V&A and implementation of adaptation options.

Additionally, international discussions about the balance between mitigation and adaptation, as well as the role of adaptation in reducing vulnerability to existing climate variability and non-climatic risks, has also been constrained by procedural and political factors. In particular, there has been a fragmentation of policy caused by the lack of a single COP agenda item to address adaptation issues, and political complications in disentangling adaptation from the conceptually distinct potential problems facing energy-exporting OPEC countries arising from implementation of response measures (Barnett et al., 2004; Yamin and Depledge, 2004).

2.1. Unexpected rapid changes and adaptation policy linkages

The Convention's adaptation provisions were agreed at a time when scientific understanding of climate change viewed it as a linear problem that would unfold gradually over the course of a century. AOSIS countries could see the threat to their existence quite clearly but many other developing countries simply regarded climate change as too distant a concern for them to give priority to climate change.

Scientific understandings suggest that climate change could be occurring more rapidly than previously thought, and we should expect abrupt change and surprises. This new understanding, combined with more detailed impact assessment work undertaken by developing countries and the realization that some climate impacts are inevitable, has shifted developing countries' perspectives. Consequently, most now give climate issues, particularly adaptation, far greater attention and urgency.

The Millennium Ecosystem Assessment and recent scientific papers under review for the IPCC Fourth Assessment Report support this prioritization, underscoring the need for policymakers to forge stronger linkages between sustainable development and anticipatory adaptation measures, particularly at a regional and sectoral level. A greater emphasis on prevention of, and preparedness for, climate-related disasters and the greater involvement of the 'disasters policy community' is also much more likely as a result of initiatives such as the January 2005 Kobe World Conference on Disaster Reduction (WCDR). The realization that the December 2004 Asian tsunami disaster might not have resulted in a quarter of a million deaths had early warning systems worked effectively has also added impetus to the need to connect climate change adaptation with other policy and institutional structures that will enhance short- and long-term coping capacity.

The increased focus on adaptation will also necessitate greater policy coherence and institutional coordination across a wide range of multilateral environmental agreements (MEAs), particularly those dealing with natural resources (such as the Convention on Biological Diversity, the Desertification Convention, and regimes dealing with wetlands, ozone, and migratory species). It will also require greater coordination with the Organization for Economic Cooperation and Development Committee on Development Assistance (OECD DAC) which is currently working on tracking funding of the Rio Conventions. Finally, because aspects of trade and finance policies are currently not aligned with the goal of human well-being and global environment protection, it will also require greater attention to policy coherence with the WTO and Bretton Woods Institutions (Yamin and Depledge, 2004).

Thus for a combination of reasons, adaptation issues will play a larger role in the international climate negotiations and other development-related negotiations in the coming decades (Burton et al., 2002; Simms et al., 2004). Post-2012 discussions will have to integrate adaptation as a core concern rather than relegating it to the back-burner, as happened in the Kyoto negotiations. These negotiations will also have to address policy and institutional coordination in an *integral, ongoing manner* that will make the Kyoto negotiations look logistically and intellectually simple!

2.2. Adaptation financing: incremental costs, mainstreaming and the GEF

The Convention's financial provisions were underpinned by expectations that human-induced climate change would be distinguishable from natural climate variability. The definition of 'climate change' in the UNFCCC reflects this and was deliberately drafted this way because Annex II Parties did not want to pay for adaptation to natural climate change.[3] Although the Third Assessment Report (TAR) confirmed that anthropogenically induced climate change is occurring, it is widely acknowledged that it is conceptually and empirically difficult to disentangle human-induced climate change from natural variability when discussing specific weather events (Stott et al., 2004). A critical issue then will be what is meant by *adaptation* and by *vulnerability*, as neither term is defined by the Convention, and what the limits to adaptation might be. Current discourse and framings, discussed below, demonstrate several meanings, and it will be important to disentangle the policy implications from different versions and nuances.

An additional complication is that the core financial provisions of the Convention deploy underlying concepts such as 'incremental costs' and 'global environmental benefits', which were designed to fund mitigation and do not fit adaptation. Furthermore, in 1992 it was expected that Official Development Assistance (ODA) trends would stabilize and improve. Instead, the last decade has seen decreases or stagnation of ODA to developing countries. Notwithstanding the Monterrey Conference on Finance in 2001 and the 2002 World Summit on Sustainable Development in Johannesburg, ODA seems unlikely to return to 1990 levels (Bezanson, 2004). A final complicating factor is that since the terrorist attacks of 11 September 2001, donors have moved to tie security concerns with developmental ones. This is skewing resources towards countries perceived to be of strategic interest to the USA as well as stretching aid budgets still further. ODA has always been subject to the donor 'flavour of the month' syndrome. But resulting unpredictability and the low ratio of aid relative to needs is one reason why developing countries remain reluctant to accept new *commitments* – even whilst most of them take *actions* over and above the baseline of legal obligation, with the result that the pace of decarbonization is appearing to increase.[4]

Although there was clear agreement that mitigation-related finances would go through the Convention's financial mechanism and GEF and would be subject to incremental cost and global environmental benefits tests, there has always been controversy about whether adaptation finances should. Current policy disputes with developing countries favouring flows through the Convention's financial mechanism and new funds established by the Marrakech Accords, and donor countries emphasizing other channels, have a long history. Article 4.4, which deals specifically with adaptation, does not refer to the concept of *incremental costs* and nor does it refer to the Convention's financial mechanism.[5] These omissions reflect disagreement about what proportion of adaptation costs would be paid by Annex II Parties and whether adaptation funding would flow primarily through the Convention's financial mechanism or through other bilateral/multilateral channels.

The issue of channels is important because the GEF is bound by its founding instrument to achieve global environmental benefits and its governance structure still gives donors more say than the COP. A major underlying issue in terms of adaptation finance is the relationship of climate funding with regular ODA. Given that many climate adaptation projects contribute to local benefits and thus make sense for a wide range of sustainable development objectives, many climate adaptation projects look like regular development projects that should be funded by the ODA, rather than the GEF. Accordingly, many donors are arguing that it is more 'efficient' for climate adaptation funding to be mainstreamed into regular ODA streams.

Currently, developing countries question such mainstreaming for several reasons. Mainstreaming adaptation funding into existing ODA flows could result in the diversion of ODA rather than 'new and additional' resources for climate change as mandated by the Convention. It could easily lead to ODA earmarked for regular development projects in health, agriculture, education and natural resource management having to deal with climate change – with the prospect of more paperwork and conditionalities for recipient countries. In any case, environmental considerations do not appear to have been fully integrated within regular development planning tools such as Poverty Reduction Strategy papers. This suggests that it may be premature to regard the use of ODA channels for climate adaptation as being more 'efficient' than dedicated channels established pursuant to the Convention (Banuri et al., 2003; Yamin, 2004). An additional concern is that, in contrast with the relative transparency of the Convention's financial mechanism, existing streams of ODA lack coordination, cannot be easily tracked, and consequently cannot be assessed for efficacy given the well-known problems relating to aid proliferation and fragmentation and limited inroads into coordinating donor programmes (Greene, 2004; Yamin and Depledge, 2004).

On the specific actions and timing of adaptation funding, developing countries have argued for many years that the 'staged approach' adopted at COP-1 should be relaxed to allow 'learning by doing' and to facilitate preparedness for climate impacts. This has been resisted by developed countries, including the EU, even though the location/context-specific nature of adaptation suggests that learning-by-doing through pilot/demonstration projects is likely to be highly effective in building capacity, pushing the boundaries of our knowledge, and highly effective in increasing public awareness. Donor insistence in earlier negotiations on making adaptation financing conditional on submission of national communications, linking adaptation funding with the issue of the submission and frequency of non-Annex I Parties' national communications, and/or creating issue linkages between adaptation funding for particularly vulnerable countries and a future action mandate have not been helpful in forging the right dynamics for the kinds of coalitions the EU will need in the coming years.

Some of these frustrations have led developing countries to consider non-ODA sources of adaptation funding. During the Protocol's negotiations, for example, developing countries devised and obtained an adaptation levy, paid by proponents of activities under the Clean Development Mechanism, which was conceptualized on the basis of ideas put forward by Brazil and AOSIS. G-77 countries wanted to extend this to the other mechanisms. During the Marrakech negotiations some EU delegations regarded ideas for generating additional funding for adaptation, such as an extended adaptation levy, in highly negative terms, stressing, for example, how the GEF and its role might be undermined. The new GEF adaptation programme with resources of around US$50–60 million is a welcome start, but this will not prove sufficient for over 130 countries' adaptation needs. Given the existing limits of donor funding and rising adaptation needs, it is likely that new ideas for adaptation funding (such as levies on offset transactions on Kyoto mechanisms in Parties and non-Parties to the Protocol) will resurface in a post-2012 context.

Long-term clarity about the amount, predictability and channels for adaptation funding, and the quality/adherence of the COP's guidance to the GEF, will be key features of future climate negotiations – as they were during the course of the UNFCCC itself. But these issues are also ones which the last three 'adaptation COPs' have failed to tackle successfully in a fundamental way. Further progress will only be made if the EU's position on adaptation funding can be backed by sound arguments and solid evidence demonstrating how developing country concerns are to be overcome in the years ahead. This will require the EU to be more constructive in rethinking about how new sources might complement, rather than undermine, existing channels.

2.3. Mitigation financing: the role of the GEF

In the medium to long term, mitigation is the most effective form of adaptation. Although the bulk of the GEF's current climate resources are spent on mitigation, not adaptation, these are still too small to make a dent in curbing GHG emissions. Thus there are also doubts about whether the Convention's financial mechanism can make a strategic enough impact with respect to mitigation.

With its limited resources, the Convention's financial mechanism can only fund a trickle of the investment needed by developing countries for the kind of low carbon pathways needed if dangerous anthropogenic interference is to be avoided (Ott et al., 2004). However, it can serve an important demonstrative and catalytic effect. Additionally, more tailor-made support, especially for large industrializing countries, needs to be provided to enable them to implement low carbon options – without such support being conditional on the acceptance of binding commitments/other regulatory instruments that do not match their specific circumstances. Because carbon markets and private finance will play a more important role in financing mitigation in some developing countries, the impact of Russian ratification on these markets, including the EU emissions allowance trading scheme, will need to be carefully evaluated to assess whether investment flows are indeed being 'greened' in developing countries and how the role of the CDM can be enhanced (Kenber, 2005).

Post-2012 discussions may therefore require a more fundamental reconsideration of the Convention's financial architecture in order to realign its stated purpose with its actual and potential impact on both sides of the mitigation/adaptation equation. Both sides of the equation will have to take into account how the climate regime intends to address the separate, difficult-to-track and largely unaccountable streams of funding flowing through bilateral/multilateral channels so as to curb their contribution to underwriting carbon-dependent pathways, through for example, subsidies and export and other guarantees. Mechanisms to assist donor coordination of climate programmes, an assessment of the impact of the Bretton Woods Institutions and other IFIs on funding carbon-dependent pathways will return to the COP agenda.[6] And as the EU Member States are the biggest supporters of the GEF, it will be incumbent on the EU to devise strategies that deal with its actual and perceived shortcomings in a constructive manner.

2.4. Adaptation in developed countries

The framing of climate change as being linear, incremental and something that in the long term might actually benefit developed countries has also had consequences for the role that developed countries have played in relation to adaptation in the climate regime. Unlike mitigation, where a wide range of economic models and assessment tools were created by developed countries for their own use because they had to reduce GHGs first, the development of vulnerability assessment and adaptation tools has tended to lag behind. This has partly to do with limited technical resources being channelled to progress work on mitigation/KP implementation. But it is also because developed countries have *assumed* that their greater resources will ensure that they can deal with adverse climate impacts and will easily be able to take advantage of the favourable potential impacts. These considerations underlie the fact that to date, few Annex I Parties' national communications contain detailed work on domestic assessment of vulnerability and implementation of adaptation options. One result is that many European countries seem unprepared for the kinds of floods, droughts and heatwaves that Europe has recently experienced.

The debate within Annex I Parties on adaptation towards a greater examination of climate impacts and adaptation strategies is shifting as public awareness of climate change within Europe increases. The realization that having the *capacity to cope* or adapt to new conditions, and even to take advantage of favourable changes, is not the same as marshalling it to achieve these ends, which is also important (Easterling et al., 2004; O'Brien et al., 2004). Work undertaken by leading governments such as the UK through its UK Climate Impacts Programme, which has pioneered greater awareness of climate impacts, and developed tools to help a wide range of actors address the necessity of adaptation in the UK, will play a major role in how Annex I Parties look at adaptation (Willows and Connell, 2003). This, in turn, will influence developing countries' approaches to formulating and implementing adaptation options even though assessment tools and adaptation choices will always have to be context-, space- and time-specific.

Unlike inventory preparation and mitigation policy, where Annex I Parties' work is highly advanced compared with comparable work done by developing countries, adaptation provides a rare example of Annex I and non-Annex I Parties working more or less in tandem on an issue which requires each Party to undertake virtually identical tasks domestically (impact assessment, adaptation option evaluation and implementation). This underscores the need to ensure coherence between, for example, what Annex I Parties are saying with regard to adaptation at home and abroad.

More significantly, it also provides tremendous opportunities for reflection and mutual learning – many of which are likely to be from South-to-South and South-to-North rather than the expected North-to-South (Pasteur, 2004). These possibilities, and insights from the organizational learning literature, thus speak for strengthening what social scientists describe as action-orientated networks working on adaptation issues that cut across geographical, political and epistemic boundaries.[7] Given existing problems such as the EU 'bunker mentality' and the additional challenges of EU enlargement, support for such networks will serve the EU well in conceiving, designing and ultimately implementing its post-2012 strategy.[8] The organizational aspects of a post-2012 EU strategy, including the development of knowledge-based coalitions, should therefore not be overlooked in favour of sound-bites and 'big ideas'.

3. External and internal dimensions of adaptation and the EU

The EU has played a strong leadership role on mitigation by, *inter alia,* supporting the legally binding nature of Kyoto, operationalizing the EU emissions trading scheme and advocating a temperature limit of 2°C as a threshold to avoid dangerous interference with the climate system. The EU Sixth Environment Action Programme also highlights the positive and constructive role the EU must play as a leading partner in the protection of the global environment and in the pursuit of a sustainable development (European Parliament, 2002, Decision No 1600/2002/Ec).

The European political response specific to climate adaptation is set out in the EU Council Conclusions of 14 October 2004 which confirms that 'worldwide societies have to prepare for and adapt to the consequences of some inevitable climate change' and that the EU reaffirms its 'continued commitment to assisting developing countries to adapt to adverse impacts of climate change' (EU Council, 2004). The basic position of the EU on assisting developing countries to adapt (i.e. how the EU views how *other* societies might adapt and the level of assistance it can provide) has been mapped out in broad terms over the last few years in international climate negotiations.

Although the EU Council talks of societies *worldwide* needing to prepare for adaptation, there has been a dearth of thinking about the EU's own adaptation strategy for climate impacts. Perhaps because of its contextual nature, the EU's adaptation agenda is fragmented in nature between external and internal elements and it is difficult to unify the different strands of various EU positions and initiatives. This section maps out the various elements focusing on development cooperation, adaptation policies for the EU, and then wider EU integration processes that might promote climate adaptation both internally and externally.

3.1. Development cooperation

The EU's *external climate policy* has focused largely on mitigation – due to domestic as well as international pressure from developing countries for Annex I Parties to take the lead in GHG mitigation, to do so domestically and to avoid overreliance on the use of carbon sinks. In the early years of the regime, tensions between Member States and the EU over subsidiarity/competence marred EU unity and masked the fact that there was little in the way of a well-defined EU programme of mitigation (Gupta and Grubb, 2000). The EU's European climate change programme of implementation since Kyoto has been keenly observed by other countries and its implementation has made the EU and Member States work more closely together, allowing the EU to present a more unified, credible and coordinated negotiating stance on mitigation.

The EU's *external stance* on adaptation lags behind its stance on mitigation in some respects. There are several reasons. As pointed out above, like other Annex I Parties, the EU has been focusing on getting the 'Kyoto' house in order and is only now beginning to grapple with the need to develop a long-term coherent framework for considering adaptation issues. The fact that the EU did not contribute to the GEF third replenishment in its own right because Member States undertook this responsibility may also be a contributory factor. On a positive note, the EU has recognized the need for adaptation, stated it will work towards a 2°C target, and has given greater priority to adaptation funding by working with other partners to deliver on the political commitment for funding, including the declaration made at COP-6 part II.[9]

To encourage Member States and the EU to integrate climate change into development cooperation, the Commission has also issued a communication on Climate Change in the Context of Development Cooperation in March 2003 and will aim to allocate additional financial resources for this purpose (European Commission, 2003). The December 2003 Council decision sets out the basic principles for incorporating adaptation into development, including through utilizing synergies with sustainable development and poverty eradication (Lamin, 2004).

An Action Plan for 2004–2008 to accompany the EU Strategy on Climate Change in the Context of Development Cooperation was agreed recently by the EU Council in October 2004. The Plan sets out core principles to guide implementation and four strategic objectives: raising the profile of climate change, support for adaptation, support for mitigation, and low GHG development paths and capacity development. Each objective is broken into sub-actions with time-frames for specified actors. A fifth section sets out monitoring and evaluation aspects of the Action Plan, which is to be continuously monitored with specific evaluation reports being prepared once every 2 years, with the first report to be ready by the end of 2006. The Plan is wide-ranging, requiring action by Member States, the EU and other European actors, including working in partnership with NGOs, international organizations including international financial institutions

(IFIs), and UNFCCC bodies working on adaptation (CGE and LEG). It sets out an impressive array of actions for each core objective – there are nearly 20 sub-actions for adaptation, with additional actions supporting capacity building and awareness. But as yet no specific additional financial resources have been created to finance the Plan. Only time and the evaluation due back in 2006 will therefore tell what progress will be achieved to this laudable but wide-ranging shopping list of ambitions.

3.2. Policies for adaptation in the EU

For the reasons described above, the EU's domestic climate policy has focused on mitigation. The first and second phases of the European Climate Change Programme (ECCP), for example, dealt with mitigation issues, to enable the EU to ratify Kyoto. Following publication of the 2004 EEA Report, the Executive Director of the EEA noted that 'strategies are needed, at European, regional, national and local level, to adapt to climate change' (EEA, 2004a). What will these strategies look like? And what methodologies, tools and information will form the basis of evaluating the range of adaptation options? How will adaptation actions within the EU be financed? What additional burdens will be picked up by individuals, governments, private sector and by EU financial institutions? What kind of information and technical support, if any, will vulnerable groups receive and from whom? And which aspects, if any, need an EU component and which are better left to Member States? Currently there appears to be no overview of how adaptation within the EU might be handled on a long-term and strategic basis. And there is little research on how European stakeholders and publics will respond to climate impacts in Europe and abroad.

The EU's third national communication to the UNFCCC reports a large number of adverse impacts (EU, 2001). It hints that policies are needed to address impacts to, *inter alia*, soils, mountains, fisheries, coasts, water management, forests, ecosystems, tourism and the need for development, e.g. regional policy, incorporation of risk management in investment decisions, and 'climate-proof' building codes and rules. But no policy measures are spelt out and no *process or time-frame* is set out to signal how those measures that might benefit from an EU input would be investigated for development at the EU level. Reasons for this might include the fact that the assessment of impacts, adaptation options and their implementation at the domestic level is not far advanced in any Member State, although some are more advanced than others, and that some of the most immediate adaptation options are in the domain of planning and development of infrastructure which Member States, not the EU, have competence.[10] Although much more work is now under way at the Member State level, the EU's external stance – that societies worldwide must adapt and that developing countries should incorporate adaptation in their national development frameworks – is thus currently not matched by its internal record on promoting adaptation within the EU.

3.3. Integration of environment and climate policies

The term 'integration' is used in the EU to refer to incorporation of environmental and social considerations into all spheres of policy-making, particularly economic policy, in much the same way that the term 'mainstreaming' is used in a development context. There is widespread recognition within the EU that environmental protection must be integrated into sectoral policies, because environmental policy alone cannot achieve the environmental improvements needed as part of sustainable development. If this were done successfully, it is doubtful whether in the long term a separate climate policy would be needed at all.

Processes to achieve environmental integration at the EU level have been set in motion by the 1999 Cardiff process with the EU Sustainable Development Strategy, adopted in Gothenburg in 2001, establishing environmental protection goals and integration alongside the EU's economic and social objectives. These processes are intended to ensure that economic, social and environmental objectives are incorporated at all levels and in all sectors. The limited progress to date, however, has led to the Commission announcing that it would carry out an annual stocktaking which will be taken into account in the preparation of its future Spring Reports, starting in 2004 (European Commission, 2004).

The important point to note here is that most of the discussions about environmental integration within the EU have focused on incorporation of GHG mitigation policies as reflected, for example, in the sectoral assessments included in the latest Commission working document (European Commission, 2004). The incorporation of climate adaptation into these broader EU processes of environmental integration has yet to be examined in any real depth.

On a positive note, the development of tools such as the Impact Assessment, scheduled to commence fully in 2004, has the potential to contribute to enhancing EU policy coherence and for integration of environmental considerations (European Commission, 2002). Impact Assessment forms part of the Better Regulation Action Plan which is designed to increase the quality, transparency and timeliness and effectiveness of EU policy interventions (Better Regulation Action Plan, 2002; Better Lawmaking, 2003). The use of Impact Assessment is encouraged in the EU and by Member States in the Action Plan to accompany the EU Strategy on Climate Change in the Context of Development Cooperation 2004–2008. A range of other tools, such as the Strategy on the Sustainable Use of Natural Resources and of Sustainability Impact Assessment to assess the impact of trade negotiations on sustainability and incorporate sustainability into EU trade policy are also under discussion (Forseback, 2001). These may encourage 'joined-up' EU thinking on vulnerability and adaptation to emerge.

These activities and initiatives demonstrate that the EU is trying to tackle seriously the challenge of integrating economic, social and environmental considerations which lie at the core of sustainable development. But they also demonstrate that its attempts to do so are at a relatively early stage of evolution. Most of the initiatives have yet to become fully operational and it will be some years before there is sufficient monitoring and evaluation experience to assess their effectiveness. In terms of devising an EU post-2012 climate strategy, these activities – and the limited, early experience of their internal and external impacts – need to feed into 'next steps' discussions. These initiatives could, for example, sensitize EU development assistance and trade policy on a sector-wide level and therefore be used to reduce climate risks for vulnerable developing countries in a way that is not possible through current project-based approaches.

4. Conclusions: A new approach to development partnerships or business as usual?

Adaptation issues provide a rare example of Annex I and non-Annex I Parties working more or less in tandem on an issue which requires each Party to undertake virtually identical tasks at the domestic level. This provides tremendous opportunities for all Parties to reflect and to engage in mutual learning. It also underscores the need to ensure coherence between, for example, what

Annex I Parties are saying with regards to how adaptation might be 'integrated' at home and how it might be 'mainstreamed' by others abroad.

The EU, like other donors, has, however, framed its stance on adaptation in terms of the 'carrots and sticks' it can provide to developing countries that need adaptation assistance. It has tended to argue that developing countries should mainstream climate consideration into development planning, whilst failing to reflect on its own early experience of integrating climate change into EU sectoral policies. This experience demonstrates that mainstreaming is not a panacea and requires resources, political will and time in order to achieve results on any significant scale. The EU's approach also fails to take into consideration the opportunities for learning and coalition-building provided by the fact that because detailed work on adaptation is only now being undertaken by developed and developing countries, many of the institutional challenges posed by integration and mainstreaming of adaptation are common to Annex I Parties and developing countries.[11] These commonalities tend to be masked, however, because donor countries tend to use funding to tell developing countries what to do, or else use funding to forge support for other policy issues. As ODA is a much less significant stream for many middle-income and larger developing countries, whose cooperation will be vital to the EU in the next steps of the climate regime, this approach is unlikely to be effective in forging the kinds of partnerships the EU needs in order to play its role as global leader more effectively.

The article suggests that the EU could reframe its stance of adaptation and on development partnerships so as to focus on forging coalitions and alliances with all those interested in *learning about the transition to sustainable development* and by providing developing countries with an honest account of the status of its internal developments in this regard. It could, for example, encourage clusters of countries (developed and developing) with similar climate challenges (e.g. sea level rise) to engage jointly in V&A capacity-building, or else devise an international policy on coastal zone management jointly. By employing a learning-based approach to its development partnerships, the EU may be able to foster more genuine cooperation and policy-relevant knowledge from the key developing country players than the donor 'sticks and carrots' approach that it has traditionally deployed.

There is no doubt that the EU has played a significant leadership role in the climate regime (Gupta and Grubb, 2000). Securing the entry into force of the Kyoto Protocol in the face of US opposition was a major achievement for the EU and for all developing countries. Maintaining this leadership will require the EU to avoid, as has sometimes happened in the past, dissonance between the internal and external dimensions of its climate policy. It will also require the EU to spend a little less time and policy attention on the USA as a negotiating partner and much more on developing countries, whose cooperation will be vital in upholding the Kyoto architecture and momentum for the climate regime as a whole until there is a change of administration, and ultimately, of policy in the USA. Focusing on deepening its partnership with developing countries over the next 4 years will ensure that the EU is in a stronger position to work towards an effective, inclusive and fair climate regime for all.

Notes

1 For a summary of future actions-related policy dialogues, see http://www.fiacc.net/.

2 Although the Kyoto Protocol provides for funding of adaptation activities through a share of the proceeds on the Clean Development Mechanism (CDM), the Convention serves as the primary source of legal obligation with respect to adaptation and is thus the focus of this article.

3 Article 1.2 defines climate change to mean change in climate that is *both* additional to natural variability and attributable to human activity.

4 One recent independent study has found that the combined impact of climate and non-climate related policies in six non-Annex I Parties over the past decade has reduced the growth of the emissions by nearly 300 million tonnes per year, i.e. only 92 million tonnes less than the 392 million tonnes of reductions mandated by Kyoto for all Annex I Parties based on the Annex I Parties emissions projection for 2010 (Pew Centre on Global Climate Change, 2002).

5 Dan Bodansky's (1994) Yale Commentary makes it clear that these omissions were deliberate.

6 COP-1 mandated the Convention secretariat to 'collect information from multilateral and regional financial institutions on activities undertaken in implementation of Article 4.1 and Article 12 of the Convention [without] introducing new forms of conditionalities' (Decision 11/CP.1). See also FCCC/TP/2003/2.

7 The Linking Climate Adaptation project (LCA project), funded by DFID and coordinated by IDS with six developing countries and IIED, for example, has established the LCA Network to provide a focal point for sharing adaptation research and policy insights. A related initiative is an EU-funded project called the BASIC project (Building and Supporting Institutional Capacity in Brazil, South Africa, India and China) which will also establish a network of expertise that cuts across North/South lines working on adaptation, mitigation and institutional issues relating to implementation/further actions under the UNFCCC and Kyoto Protocol (see http://www.ids.ac.uk/ids/env/climatechange/ for details).

8 The October 2004 Action Plan mentions the objective of establishing 'a network of expertise on climate change/MEAs/environment in Commission Headquarters and Delegations, Member States and partner countries, as outlined in the EU Strategy on Climate Change in the Context of Development Cooperation' (Sub-action 1.2.3). On adaptation, it talks of strengthening the in-country processes of consultation in which civil society organizations are engaged.

9 The joint political declaration made at COP-7 by the EU and its Member States, together with Canada, Iceland, New Zealand, Norway and Switzerland, announces their preparedness to contribute collectively €450million/US$410 million (at July 2001 exchange rates) annually by 2005 to climate change, with this level to be reviewed by 2008. The EU Council of October 2004 says that the EU is determined to provide US$ 369 million with Member States reporting their contributions in national communications as of 2006.

10 A number of tools and methodologies are now being developed. See, for example, the development of a screening tool by the World Bank to assist identification of adaptation components in development projects.

11 This type of leadership is called *structural leadership*, as it involves the use of economic and material power. Other types of leadership include *directional leadership*, which refers to leadership through demonstration (e.g. domestic implementation), and *instrumental leadership*, which refers to forging smart coalition-building, e.g. through issue linkages (see Gupta and Ringius, 2001; also Gupta and Grubb, 2000).

References

Banuri, T., Spanger-Siegfried, E., Odeh, N., 2003. The European Union's Development Policy. Think Piece, 4. IIED, London.

Barnett, J., Dessai, S., Webber, M., 2004. Will OPEC lose from the Kyoto Protocol? Energy Policy 32, 2077–2088.

Better Lawmaking, 2003. Report from the Commission on the Application of the Principles of Subsidiarity and Proportionality. COM (2003) 770 final.

Better Regulation Action Plan, 2002. COM (2002) 278 final.

Bezanson, K., 2004. Rethinking Development: The Challenge for International Development Organizations. IDS Bulletin: Climate Change and Development, (July 2004).

Bodansky, D., 1994. The United Nations Framework Convention on Climate Change: a commentary. Yale Journal of International Law 18(2).

Burton, I., Huq, S., Lim, B., Pilifosova, O., Schipper, E.L., 2002. From impacts assessment to adaptation priorities: the shaping of adaptation policy. Climate Policy 2, 145–159.

COP-10, 2004. Summary of the COP tenth session, Buenos Aires, Argentina, 6–18 December, 2004. Earth Negotiations Bulletin 12(260).

Easterling, W., Hurd, B., Smith, J., 2004. Coping with Global Climate Change: The Role of Adaptation in the United States. Pew Centre on Global Climate Change, Washington, DC.

EEA [European Environment Agency], 2004a. Impacts of Europe's Changing Climate: An Indicator-based Assessment. EEA Report No 2/2004.

EEA [European Environment Agency], 2004b. Press Release, 19 August 2004. Integrating environmental considerations into other policy areas – a stocktaking of the Cardiff process, Com (2004) 394 final. 1 June 2004.

EU [European Union], 2001. Third National Communication, 30 November 2001, Brussels, 20.12.2001, SEC (2001), 2053. European Union, Brussels.

EU Council, 2004. Conclusions, 14 October, 2004, paragraph 1. European Union, Brussels.

European Commission, 2002. Communication from the Commission on Impact Assessment. COM (2002) 276 final.

European Commission, 2003. Climate Change in the Context of Development Cooperation. COM (2003) 85 final and EU Council Decision of 1 December 2003, DEVGEN 154, ENV 656.

European Commission, 2004. Communication from the Commission COM (2004) 394 final.

European Parliament, 2002. Decision No 1600/2002/Ec of the European Parliament and of the Council of 22 July 2002.

Forseback, L., 2001. The Knowledge Economy and Climate Change, An overview of New Opportunities, Report prepared by the Swedish Delegation for Sustainable Technology.

Greene W., 2004. IDS Bulletin: Climate Change and Development.

Gupta, J. Grubb, M., (Eds), 2000. Climate Change and European Leadership: A Sustainable Role for Europe? Kluwer, Dordrecht, The Netherlands.

Gupta, J., Ringius, L., 2001. The EU's Climate Leadership: Reconciling Ambition and Reality, International Environmental Agreement, 1.

IPCC, 2001. Third Assessment Report. Cambridge University Press.

Kenber, M., 2005. In: Yamin, F. (Ed.), Climate Change and Carbon Markets: A Handbook of Emissions Reductions Worldwide. Earthscan.

Lamin, M., 2004. Climate Change and Development: The role of EU Development Cooperation. IDS Bulletin.

Linnerooth-Bayer, J., Amendola A. (Eds), 2003. Special Edition on Flood Risks in Europe. Risk Analysis 23, 537–627.

O'Brien, K., Eriksen, S., Schjolden, A., Nygaard, L.P., 2004. What's in a word? Conflicting interpretations of vulnerability in climate change research. CICERO Working Paper.

Ott, K., Klepper, G., Lingner, S., Schäfer, A., Scheffran, J., Sprinz, D., Schröder, M., 2004. Reasoning goals of climate protection: specification of Article 2 UNFCCC. German Federal Environmental Agency, Report 01/04. FEA, Berlin.

Pachauri, R.K., 2004. Climate change and its implications for development: the role of IPCC assessments. IDS Bulletin: Climate Change and Development 35(3).

Pasteur, K., 2004. Learning for development: a literature review. Lessons for change in Policy and Organizations 6, IDS, available www.ids.ac.uk.

Pew Centre on Global Climate Change, 2002. Climate Change Mitigation in Developing Countries, Brazil, China, India, Mexico, South Africa and Turkey (October 2002), page iii.

Poumadere, M., Le Mer, S., 2004. Dangerous Climate Here and Now: The 2003 Heat Wave in France, draft working paper presented to International Workshop on Dangerous Climate Change, Norwich 28–29 June 2004.

Simms, A., McGrath, J., Reid, H., 2004. Up in Smoke, NEF/IIED.

Sperling, F. (Ed.), 2003. Poverty and Climate Change: Reducing the Vulnerability of the Poor through Adaptation. Multi-agency Report.

Stott, P.A., Stone, D.A., Allen, M.R., 2004. Human contribution to the European heatwave of 2003. Nature 432, 610–614.

Verheyen, R., 2002. Adaptation to the impacts of anthropogenic climate change: the international legal framework. Review of European Community and International Environmental Law 11(2).

Willows, Connell (Eds), 2003. UKCIP, Risk, Uncertainty and Decision-making.

World Disasters Report, 2004. International Federation of the Red Cross and Red Crescent Societies.

Yamin, F., 2004. Overview. IDS Bulletin: Climate Change and Development.

Yamin, F., Depledge, J., 2004. The International Climate Change Regime: A Guide to Rules, Institutions and Procedures. Cambridge University Press.

Climate Policy 5 (2005) 363–376

Towards climate policy integration in the EU: evolving dilemmas and opportunities

Måns Nilsson[1]*, Lars J. Nilsson[2]

[1] Stockholm Environment Institute (SEI), Box 2142, 103 14 Stockholm, Sweden
[2] Environmental and Energy Systems Studies, Department of Technology and Society, Lund University, Gerdagatan 13, 223 62 Lund, Sweden

Received 10 January 2005; received in revised form 10 March 2005; accepted 14 March 2005

Abstract

Europe has positioned itself as a front runner for climate mitigation policy globally. However, to reach climate mitigation targets that go far beyond Kyoto commitments, climate policy must become more integrated with sectoral policies such as energy, transport and agriculture. To achieve such policy integration, policy reframing and dilemma sharing among sectoral actors are important mechanisms. This article proposes the basic elements of an integrated policy agenda based on an assessment of achievements to date in three sectors as well as an outlook for the future of some overarching framework conditions. The analysis of existing and emergent dilemmas and opportunities suggests that a policy agenda must be vigorously pursued internally within the EU as well as with its neighbours and the outside world. Basic elements of this agenda include innovation policies, structural change pressures, and a concerted and coherent international policy agenda where trade policy, foreign policy, climate policy and agricultural policy are working in the same direction.

Keywords: Climate; Policy; Integration; Mitigation; EU; Dilemma; Energy; Reframing; Frames

1. Introduction

European politicians have agreed that the European Union (EU) should assume a global leadership in climate change mitigation. The EU has no doubt played an important and often leading role in efforts around international agreements. With an emission trading scheme (ETS) and several other EU and national policies in place or planned, it has positioned itself in the forefront in the implementation of the UNFCCC (United Nations Framework Convention on Climate Change). The EU's Kyoto commitment is to reduce by 8% the six major greenhouse gases in 2008–2012 (EEA, 2002b). However, efforts to date have only reached −3% and greater efforts must be made to reach the Kyoto target (EC, 2005). Furthermore, Kyoto is only a first step towards the deep emission reductions needed for stabilizing atmospheric concentrations at levels around 550 ppm

---

* Corresponding author. Tel.: +46-8-4121415; fax +46-8-7230348
E-mail address: mans.nilsson@sei.se

CO_2-equivalents. Such reductions will imply fundamental changes in how sectors function, and tangible costs for many actors.

Climate policy, in particular moving beyond Kyoto, has profound implications for the development in several policy sectors including, for example, energy, transport, land use, and economic policy. The most profound climate change implications arise from decisions taken within these sectors rather than from environmental policy processes. Therefore, policy integration of climate issues into sectoral policy is a prerequisite for moving ahead with mitigation efforts. Climate policy integration is seen here as a subset of, and an analytical entry point to, the larger environmental policy integration challenge, although arguably one of overriding importance. Environmental policy integration (EPI) in general, as a political principle, has a strong constitutional backing, with Article 6 of the European Treaty reading:

> environmental protection requirements must be integrated into the definition and implementation of the Community policies […] in particular with a view to promoting sustainable development.

However, there are competing interpretations of what policy integration means in practice. Some view it as a normative prerequisite that gives absolute priority to environmental goals over other policy goals (Lafferty and Hovden, 2003), whereas others see it as the provision of institutional measures for more rational decision-making, enhancing administrative coordination and the supply of environmental knowledge (Jacob and Volkery, 2004). Indeed, it has been convincingly argued that the disappointing EPI outcomes to date relate to a lack of attention to 'micro' processes of policy formulation and implementation in administrative settings at EU and national levels (Jordan et al., 2004). Because of this, coordination mechanisms such as policy appraisal, interdepartmental task forces, and reporting requirements are advocated and used in order to enhance EPI. While recognizing the importance of such measures, we view them here mainly as a means toward the end of policy integration. Rather, we consider policy integration as a policy learning and reframing process, whereby sustainability aspects become a natural part of how actors in an economic policy sector view and understand the problems, goals, strategies and activities of the sector (Jachtenfuchs, 1997; Nilsson and Persson, 2003; Nilsson, 2005).

Actors use framing processes to interpret problems, to make them coherent with their belief systems, and to provide guideposts for analysis and action. Framing leads to different views on problems and how to solve them, and one can usually find differential frames underlying most policy controversies (Rein and Schön, 1993). To advance policy agendas, frames must be confronted, reflected upon and revisited in interactions among actors with different ideas and interests. As Peters (1998) has shown, bureaucratic politics and administrative routines challenge this process. Therefore, functioning coordination mechanisms are necessary for reframing and policy integration. However, they are hardly a sufficient condition. Of great importance, but often less considered, is what these mechanisms should be substantively concerned with. In line with recent findings in political process management literature, we subscribe to the notion that explicitly addressing *policy dilemmas* is necessary for facilitating interactions, and hence for reframing (de Bruijn et al., 2002). Policy dilemmas appear when decision-makers face connected problems, and when addressing one problem causes the amplification of other important problems. Thus, they constitute decision situations in the face of goal conflicts. Contrary to mainstream perceptions, the open recognition of dilemmas can lead to reduced conflict. Framing an issue as a dilemma means that actors find that their views and perspectives are respected and placed on the table. At the same time, it prompts actors to reflect on their own problem perceptions and recognize other actors' perspectives on the issue at hand. To advance such a process, policy analysis of climate mitigation action and the development of a strategy for combating climate change must include an analysis of the different concerns that different actors and regions have, and seek to develop new agendas that seriously address these concerns.

Alas, the two main 'traditions' of analysis of climate change mitigation both miss this analysis by deploying highly aggregated goal functions and not taking into account the distribution of impacts between regions, sectors, and social groups. One has focused on 'top-down' economic analysis that compares the costs of mitigation measures – often as reductions in GDP growth potential modelled in general equilibrium models – to their benefits, in terms of reduced damages on social and ecological systems (Nordhaus and Boyer, 2000). With this perspective, costs of mitigation, overall, seem to outweigh benefits, although the long-term relative effects on gross world product are modest. The other approach has focused on 'bottom-up' techno-economic analyses, including inventories of policy options that minimize the trade-offs and build on the synergies between, for example, economic and climate mitigation goals. With this perspective, the opportunities seem ample for action, through energy efficiency measures, and the application of renewable energy technologies such as wind, solar and biomass (Azar and Schneider, 2002; Pacala and Socolow, 2004). These strands of analysis depart from completely different analytical perspectives and causal assumptions, and therefore lead to largely incomparable results. As a consequence, competing actors usually do not accept each other's arguments, leading to 'dialogues of the deaf' (Van Eeten, 1999).

It is well known among political scientists that goal conflicts and distributional impacts are key constraints for policy (Lenschow, 2003). Interest groups block or slow down reforms despite overwhelming evidence and consensus among analysts, policymakers, and in many cases also the public, about the necessity for reform in, for instance, the Common Agricultural Policy (CAP), EU fisheries policies, national pension systems, and energy markets. In climate policy, even efforts made to reach the relatively modest Kyoto target of −8% meet strong resistance from European industries. Deeper emission cuts will have much stronger impacts on several sectors. This means that sectoral concerns must be addressed more strategically if climate mitigation policy is to advance. This is also recognized in EU's stocktaking:

> ... a difficult process: as many of the 'low hanging fruits' of integration have already been picked, future efforts to reverse persisting unsustainable trends will need to focus increasingly on structural reforms, which may generate tensions with established interest groups in the sectors concerned (EC, 2004a, p. 3).

At the forefront of politicians' and interest groups' attention are several goal conflicts and related policy dilemmas between climate mitigation and sectoral policy goals.

1.1. Objective and structure

The overall objective of this article is to identify key challenges and opportunities for climate policy integration in European sectoral policy areas. It focuses the analysis on important policy dilemmas in Europe that result from present and future efforts to integrate climate mitigation in sector policies. These dilemmas are currently barriers to climate policy integration, and inhibit the effective functioning of policy coordination mechanisms. However, analysing the coordination mechanisms themselves is beyond the scope of this article. It should be noted that the policy dilemmas are nothing new; rather they reflect well-known goal conflicts based on the substantive positions and interests of different stakeholders. The novelty is to address them in the context of an integrated future policy outlook. The underlying assumption is that deep emission reductions far beyond the Kyoto targets are necessary in the next 30–50 years. Even with abundant 'no-regrets' mitigation options, such ambitious reductions are likely to challenge key interests of powerful actors. The analysis is based on an assessment, at the European policy level, of the climate policy integration achieved so far in three important policy sectors; energy, transport and agriculture; as well as an outlook of some trends that will have important implications on Europe's future prospects to undertake advanced climate strategies while safeguarding its welfare goals. In the discussion we proceed to highlight evolving key policy dilemmas and issues that need to be addressed. Finally, based on this discussion, some policy implications are drawn.

2. Where we are today – achievements so far

In 2002, the EU-15 was a little more than one-third of the way (–2.9%) towards achieving the Kyoto target of –8% (EC, 2005). A recent assessment of greenhouse gas emission trends and projections in Europe concludes that existing domestic policies and measures will reduce total EU-15 emissions by only 1% from base-year levels by 2010 (see Table 1). When additional domestic policies and measures being planned by Member States are taken into account, an emissions reduction of 7.7% is projected (EEA, 2004). This, together with the use of Kyoto mechanisms, means that the Kyoto target of –8% could be achieved. It assumes, however, over-delivery by several Member States (Finland, France, Greece, Ireland, Sweden and the UK) compared with the burden-sharing targets.

Table 1. Changes in EU-15 greenhouse gas emissions by sector (EEA, 2004)

	Share of total emissions in 2002	Past emissions 1990–2002	Projections for 2010 with existing measures	Projections for 2010 with additional measures
Energy excl. transport	61%	–5%	–3%	–11%
Transport	21%	+22%	+34%	+22%
Industrial processes	6%	–22%	–6%	–26%
Agriculture	10%	–9%	–13%	–15%
Waste	2%	–27%	–54%	–58%
Total	100%	–2.9%	–1.0%	–7.7%

EU's strategic goals and accompanying policies and measures demonstrate that it is on the way to establishing political preconditions for a sustainable development policy which takes climate mitigation seriously. Many of them will only start to deliver in future years, and further proposals are currently under development. Important common policies and measures include, among others, directives on carbon trading, electricity from renewable sources, combined heat and power, building-energy performance, biofuels, waste, and energy crops. In 1999, the **Cardiff process** was launched to promote the integration of environmental issues into sectoral policy. A recent stocktaking reports that progress has been slow, there has been significant variation between sectors, and political commitment is often too weak. It notes that policy measures that lead to the internalization of environmental costs offer one of the fastest routes for informing the decisions of both economic operators and policymakers in the concerned sectors. However, competence in this field principally lies with the Member States (EC, 2004a).

In 2000, the European Commission launched the **European Climate Change Programme** (ECCP) with the goal of identifying and developing all the necessary elements of a strategy to implement the Kyoto Protocol. Through a cooperative effort of relevant stakeholders, the purpose of the ECCP was to undertake preparatory work for the Commission to propose in due course concrete policy proposals to the Council and the European Parliament. The ECCP has been an important force in the complex process of EU policy-making and reportedly it has also promoted the issue of horizontal integration of climate policy across the Directorates General of the Commission (EC, 2001c).

The **Lisbon strategy** has been since 2000 the central instrument guiding political action in the EU. At the Lisbon European Council in March 2000 the EU set a strategic goal for the next decade

of becoming the most competitive and dynamic knowledge-based economy in the world capable of sustainable economic growth with more and better jobs and greater social cohesion.

In 2001, the Gothenburg European Council adopted a **Sustainable Development Strategy** with the addition of an environmental part to the Lisbon Strategy. A set of 'structural indicators' is used for monitoring progress towards the Lisbon goal in each of the Member States, covering five domains: employment, innovation and research, economic reform, social cohesion, environment, as well as general economic background (EC, 2004b).

It is often relatively easy to agree on general policy goals and principles; for example, sustainable development and the importance of integrating environmental issues into sectoral policy. However, this does not mean that the implementation process then proceeds without problems. At the sectoral level various dilemmas surface, and the original goals or intentions are abandoned or diluted. Important sectors in the context of climate change include energy, transport and agriculture.

2.1. The energy sector

Energy industries alone cause around 30% of overall greenhouse gas emissions in Europe (EEA, 2003). In addition, there are energy-related emissions from other industries and sectors so that total emissions from energy, excluding transport, account for 61% of total greenhouse gas emissions. The proportion of renewable energy sources in total energy supply has slowly increased since 1992, and by 2001 the share of renewables had reached 6% overall, compared with 40% for oil, 23% for natural gas, 15% for solid fuels and 16% for nuclear power. The share of renewables in electricity production was 13% (EEA, 2002a).

In contrast with agriculture, for instance, there is no basis for a common energy policy in the EU, as Member States have wanted to maintain competence. Nevertheless, with the internal market or environmental protection as a legal basis, important policy has been developed for energy. The most important developments are the electricity and gas market directives in 1996 and 1998. Although the energy market reform process is still ongoing in most countries, it has changed profoundly what policy measures are available for climate change mitigation and other environmental protection. The aim of market reform was to achieve market integration and lower prices for industrial competitiveness, and there were few, if any, concerns about climate. Instead, climate has been integrated in the EU's energy programmes ever since the early 1990s. In 1992, the commission presented a climate change strategy, including an efficiency programme (SAVE) and a programme for renewable energy (ALTENER). In later years, initiatives have accelerated. In 2001, a renewable electricity directive set an indicative production target of 22% (compared with 13% today). In 2002, an emissions trading directive was adopted, covering 45% of total emissions and most of the energy sector emissions. Other relevant initiatives include the directives on labelling, building energy performance, and combined heat and power, as well as the proposed directive on energy services.

Thus, climate policy integration in the energy sector has been relatively successful; on paper at least. However, fundamental technology shifts are needed for ambitious emission reductions. Specifically, as shown in low-emission scenarios, this entails high rates of energy intensity improvements, sustained over long time periods, combined with a shift to carbon-neutral energy sources (i.e. renewables and nuclear), and technologies for carbon removal and storage. Despite the rhetoric about win–win measures, it is likely that moving beyond the Kyoto commitments will be associated with permanently higher energy prices, particularly for industry. This poses a dilemma, as low energy prices have been a cornerstone in national energy policies to date, with industry typically facing lower energy taxes than other parts of society. Already the emissions trading scheme (ETS) that is currently put in place has been framed as threatening basic materials industries such as mining, iron and steel, and paper and pulp, as well as engineering and manufacturing industries with high-energy input (Markussen and Tinggard Svendsen, 2005). National allocation plans under the ETS have been generous, showing that politicians take this dilemma very seriously (Betz et al., 2004). The ETS also exemplifies how a policy instrument goes through changes in the implementation process so that in its final detailed design it produces some awkward incentives (Johansson, 2005). So far, European industry has not been very interested in constructively addressing the dilemma of industrial competitiveness versus internalizing climate costs. The risk of carbon leakage has been used as an argument against carbon taxes or emissions trading. This argument can be debated, but the political constraints for moving forward are evidence enough that it must be dealt with. The competitiveness of European industries is clearly framed as contingent on maintaining a level playing field internationally in terms of costs of energy and climate mitigation.

2.2. The transport sector

Transport contributes to more than 20% of overall greenhouse gas emissions. It is the main area of growth in emissions in Europe, having increased by 22% between 1990 and 2002 (EEA, 2004). Specifically, there has been strong growth in road transport where aggregate volumes of person-km and tonne-km have more than doubled and tripled, respectively, in the past 30 years, and trends persist. In the last decade, road transport has come to dominate freight, and railroads have lost major shares. Road transport accounts for about 85% of all transport carbon dioxide emissions.

Although provided for already in the Treaty of Rome, a common transport policy did not start to develop until after 1985. Its guiding principle over the past 10 years or so has been the opening of the transport market (EC, 2001a). There is yet no common vision or focal point shared between governments, car-makers and fuel suppliers concerning the future combination of car technology and new fuels, despite 30 years of concern over oil dependence in the transport sector. The growth in carbon dioxide emissions may be somewhat dampened by the negotiated agreement with car-makers to improve fuel efficiency, and the Directive 2003/30/EC on the promotion of biofuels with an indicative target of 5.75% biofuels contribution to consumption by 2010. However, for reducing total emissions, further efficiency improvements and fuel switching are necessary, as is breaking the strong link between economic growth and growth in road transport and related emissions. Bigger cars and more powerful engines have offset impressive improvements in fuel efficiency during the past 30 years. The hydrogen-based transport system remains a relatively distant vision – one that for technical and economic reasons is hotly debated (Service, 2004). In the meantime, environmentally questionable alternatives such as ethanol and rape methyl ester (RME) from agricultural crops are being pursued, and strong policies to promote fuel efficiency in new cars are lacking.

The obvious measure of increasing the cost of road transport through higher fuel/vehicle taxes or road/congestion charges has proved to be difficult to sell to the public as well as to industry. Increasing the cost of road transport and revitalizing other modes of transport may produce logistic and engine efficiency improvements, and cause some modal shifts, but a trend-break seems a very distant prospect. Still, the cost of transport typically represents only 2–4% of the sales value of products, with fuel costs representing an even smaller share (Åhman, 2004). These costs are easily dwarfed by the production cost reductions that can be achieved through the ongoing specialization and relocation of production facilities, not least to the new Member States.

Thus, climate policy integration in the transport sector is, at best, a mixed story. Technical improvements and alternative fuels aside, increasing mobility and accessibility for economic integration and growth on the one hand, and limiting detrimental impacts from growing road transport on the other, remains a dilemma. Mobility and accessibility at low cost remains the main priority and is seen as key to European growth and integration prospects (EC, 1996a). Although a number of measures to stimulate modal shifts away from road transport are proposed in the White Paper on transport policy, there is also a warning that 'To take drastic action to shift the balance between modes – even if it were possible – could very well destabilise the whole transport system and have negative repercussions on the candidate countries' (EC, 2001a, p. 14). Thus, the White Paper is careful to balance the Gothenburg European Council's sustainable development strategy, which puts particular emphasis on the need for shifting the balance between modes of transport.

2.3. The agricultural sector

Agriculture accounts for about 10% of overall greenhouse gas emissions, primarily from methane and nitrous oxide emissions (EEA, 2003). In addition to reducing the sector's own emissions, agriculture has a potentially major role to play in mitigation through bioenergy production, creating a strong link between energy and agriculture. In 1997, a White Paper called for an increase in the share of renewables in primary energy supply from 6% (1996) to 12% in 2010. A possible strategy suggested by the Commission is to triple the use of biomass energy (EC, 1996b). Reaching this target would require 1–2 EJ (exajouks) per year of energy crops, corresponding to 5–10 million hectares of land, assuming annual yields of 10 t/ha.

The general objectives for agricultural policy (i.e. increasing agricultural productivity, ensuring a reasonable standard of living for farm workers, stabilizing markets, securing supply, and guaranteeing fair prices for the consumer) were already set in the Treaty of Rome, signed in 1957. Environmental protection was first acknowledged as an important objective in the major CAP reform of 1992 and further recognized in the following round of reform under Agenda 2000, where rural development, including agro-environment support, was given the status of the 'second pillar' of the CAP.[1] Environmental policy integration in the agricultural sector has been concerned mainly with preventing pollution (to air, water and soil) and preservation of natural heritage. Climate change and the implications of renewable energy targets for agricultural policy have been low on the agenda. It was not until the 2003 CAP mid-term review that a subsidy was introduced for energy crops on agricultural land, limited to 1.5 million hectares.

Further efforts in agriculture have concentrated on ethanol from sugar and starch-rich annual crops, and bio-diesel from rapeseed. This may seem a comfortable strategy from a traditional agricultural sector perspective, but any serious comparison will show that this is clearly environmentally inferior to cultivating cellulosic or herbaceous perennial crops. The Working Group on Agriculture under the ECCP tackled the mitigation potential of biomass for energy/industry as one issue, together with reducing nitrous oxide emissions from soils, carbon sequestration potentials, methane emissions from enteric fermentation, and methane and nitrous oxide emissions from manure (EC, 2001b). It was also noted that activities in the agricultural sector depend on actions in other sectors, e.g. energy and transport, and that a more coordinated approach is needed.

Agricultural policy has been slow to react to the climate change issue or other energy and transport policy drivers (i.e. energy security concerns) working in the same direction. Paradoxically, agriculture is the area alongside internal markets where EU has exclusive competence and, in theory, could have an impact if there was agreement. Reaching targets for biofuels in transport, renewable electricity, and renewable energy means moving into new patterns of land use, with tens of millions of hectares planted with energy crops. This poses important real and perceived policy dilemmas for decision-makers in an area already fraught with conflicting goals between economic efficiency and the protection of rural employment, landscapes, and culture. Implementation will require considerable changes in agricultural policy that will challenge vested interests, powerful lobbies, and cultural views and traditions around Europe. At the same time, the agricultural sector holds the key to reforming the global trade regime and the enlargement of the EU makes it impossible to uphold current support levels in the long term. Beyond 2010 there may be considerable competition for land between the new demands for bioenergy, renewable chemicals, feedstock production, and traditional food production. This competition may be exacerbated by the extensive farming practices currently discussed under the CAP.

3. Where we are going – a changing context

Forward-looking policy analysis recognizes that the framework conditions for climate mitigation policy will change in the long run, in ways that will often be unpredictable, but in some cases might be predicted or even influenced. The nexus of actors–powers–interests that constrains policy action today may transform itself many times, as may price relations, geopolitical relations, and international regimes. The following sections describe a first and limited set of factors that constitute part of the framework conditions for addressing climate mitigation dilemmas. Such factors will modify how the dilemmas are framed in policy deliberations, and as a result also the political potential for climate mitigation beyond Kyoto commitments.

3.1. Fossil energy supply

Oil and gas dependence is already causing increasing strategic concern for Europe. The political and social structures in the world's leading oil producers, most notably Saudi Arabia, may lead to collapses in these societies (The Economist, 2004). Furthermore, the situation between the Arab States and the West is strained. Oil prices react to this, as we have seen during 2004 and 2005. On the gas side, the conflict between the Russian government and their 'oligarchs' may threaten natural gas supply. Given the possibly calamitous political scenarios, a long-term strategic reliance on these regions and countries does not appear to be an attractive option. Giving supply security issues a greater weight in the equation may fundamentally reshape how costs and benefits of climate mitigation are understood, also in a relatively short-term scenario. Furthermore, although we cannot assume to know what market prices will be, US$40–50 per barrel as a potential price range is, on balance, not an unreasonable estimate. In this range, renewable technologies, synthetic fuels, and unconventional oil reserves become viable within and outside Europe, thus enhancing security of supply.[2] It should be noted that although there will be consumer responses to fuel prices in this price range, it is unlikely, on its own, to provoke any trend-breaks in the development of, for example, road transport.

3.2. Global climate negotiations

Climate policy, from the Villach conference in 1985 to the present international regime, has developed at remarkable speed. Clearly, a lot could happen in the next 20 years as well, especially if the EU pushes negotiations forward. However, while the Kyoto Protocol has been an important start, it is unlikely that the global UNFCCC process or its successor would yield strong mitigation commitments (towards 550 ppm CO_2-eq.) in the future. Global agreement is desirable but it is also likely that the EU will have ambitions that go beyond the 'least common denominator' commitments that global negotiation processes tend to conclude in.[3] It is conceivable that climate policy may regionalize instead, along with other policy areas such as trade. European nations seem better positioned than most other regions to take this further. For instance, the political climate in the USA makes an advanced climate policy hard to conceive of at present, for political as well as for structural reasons. The access of industrial interest groups to political decision-making has increased and the general public is blissfully ignorant of foreign affairs (Roman, 2004). For the EU, an advanced regional strategy might therefore become an increasingly attractive alternative to global inertia.

3.3. Global trade regimes

WTO-led trade liberalization has had difficulties advancing. The EU and the USA have been reluctant to reduce the protection of their agricultural sectors, to the profound frustration of the South. At the same time, it has not been possible to include standards for production, including labour and environmental standards, in the global trade regime, creating concern among Northern interest groups and governments. Consequently, the 'Doha' round of trade talks has more or less ground to a halt a few times already. In order to revamp the international trade regime and re-institute public trust in the WTO, existing incompatibilities between social, developmental, and environmental policy and international trade regimes need to be explicitly addressed. One can realistically project that these aspects will lead to major opportunities for change in trade regimes in the long run. For instance, this could entail that production processes abroad become susceptible to regulation, and that border adjustment taxes are imposed on traded goods (National Board of Trade, 2004). Such schemes would offset possible competitiveness losses suffered by European industries subject to stricter environmental and social regimes.

3.4. Enlargement of the EU

As the french and Dutch referenda in May 2005 suggested, deeper political integration and more common resources to spend on sectors and policies in Europe is not necessarily the future for the EU. A broader and more diversified EU with more low-income countries might spawn reluctance among Member States to cede more competencies to Brussels. Rather than centralizing more powers, the EU might move towards a policy regime of cooperating nations. This, however, does not mean that the EU cannot demonstrate leadership. In fact, in areas where the EU has had centralized competency (i.e. the CAP), global leadership has been remarkably poor whereas leadership has been stronger in areas where progressive Member States can advance the agenda, such as climate policy. Enlargement will also, in addition to WTO pressures, lead to further reforms of the CAP, which carries with it a potential to enhance the contribution to climate mitigation. Furthermore, the EU must handle its relations towards new countries on its eastern and southern borders, assuming that 'Fortress Europe' is neither desirable nor sustainable.

4. Policy implications

Contextual developments concerning energy resources, climate negotiations, trade regimes, or EU enlargement, do not fundamentally diminish the importance of current climate policy dilemmas. However, they might present new opportunities for addressing them. Correspondingly, climate policy can motivate efforts to influence their development. In the following, we discuss the implications of this for how the EU could proceed internally in the context of implementing the Lisbon strategy as well as on the international policy arena in the context of formulating a coherent foreign policy agenda.

4.1. Opportunities in the context of the Lisbon strategy

Voluntary approaches and win–win opportunities aside, climate mitigation will need to entail a stronger policy pressure on prices through taxes or emissions trading. A first 'toddler' step has been taken through the emissions trading scheme, but further steps are necessary to drive technology and structural change. The Lisbon strategic goal to become a knowledge-based economy suggests a strong momentum for such change. In many ways, it resonates with the climate change agenda. However, this requires a reframing about what it is that drives competitiveness in the long term. One could argue that the traditional energy policy objective of inexpensive energy in order to support competitiveness in basic materials industries is not in agreement with the Lisbon vision of a knowledge-based economy. However, the political courage, leadership, and responsibility for Europe as a whole, and the capacity to pursue long-term visions necessary for reaching the Lisbon goals, seem to be lacking today. Powerful economic and political actors have vested interests in the *status quo* and will be impacted by a transition towards the Lisbon vision.

Structural change is always ongoing and a natural part of business life, and industry has often exaggerated fears of the costs of adjusting to environmental regulations (Stockholm Environment Institute, 1999). As noted above, with a higher energy price range, new technologies may become viable relatively quickly. Nonetheless, change must be managed and steered in the right direction. Competitiveness losses might jeopardize the innovation capacity needed for supporting the technology shift. If Europe, rather than its competitors, is to supply these new technologies, then innovation policies and governance systems to promote technology development and market introduction will need to accompany the economic pressure. Traditional innovation policies often take a narrow focus on R&D support, although technology change is a complex evolutionary process. Furthermore, different technologies have different characteristics and involve different types of actors (producers, suppliers, buyers, users, etc.) in various organizational and institutional settings. Reducing stand-by losses in appliances is a different ballgame from nurturing markets for wind turbines, which in turn is different from improving energy efficiency in buildings or implementing systems for carbon capture and storage. But in all cases there is a role for governments, at various levels, to manage the transitions to new solutions (Kemp et al., 1998). A comprehensive governance of technological change might entail, for instance, public investment in downstream adoption and learning, public–private partnerships, supporting arenas for niches to develop and grow, and sustained political support, over long time periods (Kok et al., 2002).

However, this active role is seemingly in conflict with the widely shared ideal of 'small' government and deregulated markets. The decoupling of economic growth from transport sector emissions is a case in point. It is plausible that the transport work will continue to increase and that modal shifts will not significantly reduce road transport. The main fuel options for the long term under strict climate restrictions are likely to be carbon-neutral electricity or hydrogen (or hydrogen carriers) in combination with new vehicle technologies. However, such new infrastructures can probably materialize only if there is a widespread consensus on strategic choices and strong government involvement. Showing leadership in climate technology therefore also means that governments cannot, in the long run, resort to merely correcting market imperfections.

4.2. Opportunities in the context of an EU foreign policy agenda

The combination of 'push-and-pull' policies in a concerted internal climate policy agenda discussed above is relatively well endorsed on paper (EC, 2005). However, less debated is how this agenda could be coordinated with external policy, although the mechanisms for such coordination have been explored recently in the face of the proposed Constitution (van Schaik and Egenhofer, 2005). The EU is still in the process of defining its foreign policy agenda and finding its geopolitical role. Foreign policy, trade policy, climate policy and agricultural policy must be better coordinated in the search for a concerted and coherent EU policy agenda. The EU's ambition for leadership in the climate issue implies that it can also become a strong and leading force for reforming trade and its interplay with agricultural and industrial policies. Looking beyond enlargement, there is potential for enhancing relationships that the enlarged Union forges with its neighbours, such as Africa, the former USSR States and the Middle East. Given the population and economic dynamics linking the EU and these regions, they should be considered EU interest spheres to a greater degree than today. Furthermore, increasing unilateralism in the USA and the polarization with many developing countries, some of which are close to Europe, suggest that Europe should play a leadership role as a bridge and facilitator for dialogue (Biermann, 2005). Enhanced economic and cultural exchange between the EU and neighbouring states can give rise to new patterns of collaboration, trade, and production on, for instance, energy, food and environmental issues, including climate. Conceivably, Europe could import less oil and more food, to the benefit of potential food-exporting countries in poorer parts of the world.

At the global level, the EU has an opportunity to act on reforming trade, climate and agricultural policies in a coherent way. Given current pressures and conflicts in the WTO systems, more ambitious reforms of the global trade regime can be pursued to enable a more instrumental role for environmental and social governance globally. At the same time, the EU must for several reasons enhance its efforts to reform and restructure agriculture domestically. The abandonment of agricultural subsidies and export support can then easily be linked with a discussion of a global trade regime that opens the way for carefully designed climate-protective policies such as carbon taxes and accompanying border adjustment taxes. Some measures might be overruled by the WTO, at least when it comes to border adjustment taxes for production processes (Chambers, 2001). However, several measures may very well be allowed already under current rules, but have not been tested yet (National Board of Trade, 2004). Furthermore, although WTO rules today tend to become the overarching ones, there is some momentum to put multilateral trade agreements and multilateral environmental agreements on an equal footing, through, for instance, overarching dispute settlement mechanisms for all international agreements.

5. Conclusions

Europe has shown leadership in climate policy and the conditions for continuing to lead are good. Relatively far-reaching policy integration has occurred in the energy sector, with instruments in place, although presently set at rather weak levels. Results in the transportation and agricultural sectors are more mixed, with climate mitigation being far from a priority. With such poor integration in the agriculture and transport sector policies, it will be difficult for the EU to maintain a global leadership role. A lack of implementation of its ambitions in all significant sectors may in the long run undermine the credibility of the European voice globally.

Existing frameworks built around the sectors' problems, purposes and strategies (and the goal conflicts with economic integration, regional development, competitiveness, and employment) constrain whether and how climate policy is taken into consideration. The policy dilemmas that arise when driving technological change and trend-breaks in production and consumption are instrumental in the policy debate. Moving forward requires that these dilemmas be dealt with more explicitly. This also signifies an opportunity. Current debates usually take a static view on policy dilemmas, although the framework conditions for addressing them depend on dynamic factors. In fact, a broader outlook points towards opportunities for Europe to develop and drive a progressive integrated policy agenda.

The EU should develop a domestic innovation policy agenda that applies pressure by economic measures for structural change across all sectors and supports the establishment of a comparative advantage in climate-friendly technologies. This combination of push and pull policies must, however, link to a coherent foreign-, trade-, and climate-policy agenda including an enhanced interaction with neighbouring states in Africa, Eastern Europe and Asia, and the pursuit of a global trade regime that is sensitive to climate as well as other environmental and social action. Taking leadership in climate mitigation seriously, the consequence for the EU could be to pursue a policy agenda whereby countries that choose to stand outside climate mitigation efforts cannot expect to benefit from an entirely free trading arrangement with Europe.

Developing such a coherent policy agenda requires enhanced policy coordination and integration across sectors. Within the EU, it is necessary to put in place institutional arrangements that break up intra sector decision making and strengthen procedures for information and resource exchanges between sectors and between external and internal policies.

Acknowledgements

This article was written with support from the European Commission through EFIEA, based on work funded by the Swedish Research Council for Environment, Agricultural Sciences and Spatial Planning (Måns Nilsson) and the Swedish Energy Agency (Lars J. Nilsson). We thank Bengt Johansson, Bo Kjellén, Petra Menander Åhman and Max Åhman; as well as the participants of the EFIEA Conference 'Towards a Long-term European Strategy on Climate Change Policy', held on 30–31 August 2004, at The Hague, The Netherlands; for valuable discussions and comments on a previous version of the article.

Notes

1 The commodity regimes under which the lion's share of subsidies has been handed out are now widely referred to as the 'first pillar' of the CAP.
2 For an overview of energy sources, see Rogner (2000).
3 For instance, many observers argued that the R10+10 Summit in Johannesburg was dominated by rhetoric and symbolic action and should be the last of the global 'mega meetings' (The Economist, 2002).

References

Azar, C., Schneider, S., 2002. Are the economic costs of stabilizing the atmosphere really prohibitive? Ecological Economics 42, 73–80.

Åhman, M.A., 2004. Closer Look at Freight Transport and Economic Growth in Sweden: Are there Any Opportunities for Decoupling? Swedish Environmental Protection Agency, Stockholm.

Betz, R., Eichhammer, W., Schleich, J., 2004. Designing national allocation plans for EU-emissions trading: a first analysis of the outcomes. Energy and Environment 15, 375–425.

Biermann, F., 2005. Between the USA and the South: strategic choices for European climate policy. Climate Policy 5, this issue.

de Bruijn, H., ten Heuvelhof, E., 't Veld, R., 2002. Process Management. Kluwer, Dordrecht, The Netherlands.

Chambers, W.B. (Ed.), 2001. Interlinkages: The Kyoto Protocol and the International Trade and Investment Regimes. United National University Press, New York.

Economist, 2002. A few green shoots. 29 August 2002 [available at http://www.economist.com/printedition/ displayStory.cfm?Story_ID=1301796].

Economist, 2004. Crisis in Saudi Arabia. 4 March 2004 [available at http://www.economist.com/displaystory .cfm?story_id=S%27%298%3C%24P1%3F%25%23%40%21%2C%0A].

EC, 1996a. Community Guidelines for the Development of the Trans-European Transport Network: 1692/96/EC. European Commission, Brussels.

EC, 1996b. Energy for the Future: Renewable Sources of Energy, COM(96)576. European Commission, Brussels.

EC, 2001a. White Paper: European Transport Policy for 2010: Time to Decide. European Commission, Brussels.

EC, 2001b. ECCP Report June 2001. European Commission, Brussels.

EC, 2001c. Final Report: Mitigation Potential of Greenhouse Gases in the Agricultural Sector ECCP Working Group 7 – Agriculture. European Commission, Brussels.

EC, 2004a. European Commission Working Document: Integrating Environmental Considerations into Other Policy Areas – A Stocktaking of the Cardiff Process, COM(2004)394. European Commission, Brussels.

EC, 2004b. Implementation of the Lisbon Strategy, COM(2004)29. European Commission, Brussels.

EC, 2005. Winning the Battle Against Climate Change, COM(2005)35. European Commission, Brussels.

EEA, 2002a. Energy and Environment in the European Union. European Environment Agency, Copenhagen.

EEA, 2002b, Analysis and Comparison of National and EU-wide Projections of Greenhouse Gas Emissions. European Environment Agency, Copenhagen.

EEA, 2003. Europe's Environment: The Third Assessment. European Environment Agency, Copenhagen.

EEA, 2004. Greenhouse Gas Emission Trends and Projections in Europe 2004. European Environment Agency, Copenhagen.

Van Eeten, M., 1999. Dialogues of the Deaf: Defining New Agendas for Environmental Deadlocks. Eburon, Delft, The Netherlands.

Jachtenfuchs, M., 1997. International Policy-Making as a Learning Process? Avebury, Aldershot, UK.

Jacob, K., Volkery, A., 2004. Do governments regulate themselves? A comparison of tools for environmental policy integration in OECD countries. Journal of Comparative Policy Analysis 6, 291–309.

Johansson, B., 2005. Climate policy instruments and industry: effects and potential responses in the Swedish context. Accepted for publication in Energy Policy.

Jordan, A., Schout, A., Zito, A., 2004. Coordinating European Union Environmental Policy: Shifting from Active to Passive Coordination? CSERGE Working Paper EDM 04-05. University of East Anglia, Norwich, UK.

Kemp, R., Schot, J., Hoogma, R., 1998. Regime shifts to sustainability through processes of niche formation: the approach of strategic niche management. Technology Analysis and Strategic Management 10, 175–195.

Kok, M.TJ., Vermeulen, W.J.V., Faaij, A.P.C., de Jager, D., 2002. Global Warming and Social Innovation. Earthscan, London.

Lafferty, W., Hovden, E., 2003. Environmental policy integration: towards an analytical framework. Environmental Politics 12, 1–22.

Lenschow, A., 2003. Greening the European Union: are there lessons to be learned for international environmental policy? Global Environmental Change 12, 241–245.

Markussen, P., Tinggard Svendsen, G., 2005. Industry lobbying and the political economy of GHG trade in the European Union. Energy Policy 33, 245–255.

National Board of Trade, 2004. Climate and Trade Rules – Harmony or Conflict? Kommerskollegium, Stockholm.

Nilsson, M., 2005. Learning, frames, and environmental policy integration: the case of Swedish energy policy. Environment and Planning C: Government and Policy 23, 207–226.

Nilsson, M., Persson, Å., 2003. Framework for analysing environmental policy integration. Journal of Environmental Policy and Planning 5, 333–359.

Nordhaus, W.D., Boyer, J., 2000. Warming the World: Economic Models of Global Warming. MIT Press, Cambridge, MA.

Pacala, S., Socolow, R., 2004. Stabilization wedges: solving the climate problem for the next 50 years with current technologies. Science 305, 968–972.

Peters, B.G., 1998. Managing horizontal government: the politics of coordination. Public Administration 76, 295–311.

Rein, M., Schön, D., 1993. Reframing policy discourse. In Fischer, F., Forester, J. (Eds), The Argumentative Turn in Policy Analysis and Planning. Duke University Press, Durham, NC, pp. 145–166.

Rogner, H.-H., 2000. Energy resources. In Goldemberg, J. (Ed.), World Energy Assessment. UNDP, UNDESA, and WEC, New York, pp. 135–166.

Roman, M., 2004. United Stances of America: Opportunities and Pitfalls in US Climate Change Policies. Stockholm School of Economics, Stockholm.

Service, R.F., 2004. The hydrogen backlash. Science 305, 958–961.

van Schaik, L., Egenhofer, C., 2005. Improving the Climate: Will the New Constitution Strengthen the EU's Performance in International Climate Negotiations? Centre for European Policy Studies, Brussels.

Stockholm Environment Institute, 1999. Costs and Strategies Presented by Industry During the Negotiations of Environmental Regulations. SEI, Stockholm.

Climate Policy 5 (2005) 377–391

Rationales for adaptation in EU climate change policies

Frans Berkhout*

Institute for Environmental Studies (IVM), Vrije Universiteit, De Boelelaan 1087, 1081 HV Amsterdam

Received 19 January 2005; received in revised form 10 March 2005; accepted 10 March 2005

Abstract

This article sets out a series of rationales for public policy related to adaptation to the impacts of climatic change in the EU. It begins by arguing that both mitigation and adaptation are necessary parts of a coordinated policy response to the problem of climatic change. However, the 'problem structure' of adaptation is significantly different from that of mitigation. For instance, adaptation may generate private benefits that are likely to be experienced over the short term, relative to benefits associated with the impacts of mitigation actions which are public and experienced over the longer term. This divergence influences public policy rationales for adaptation and poses challenges for the integration of mitigation and adaptation in climate policies. Five key challenges facing climate adaptation are identified, and these are used as a basis for proposing rationales for policy action on climate adaptation. These relate to: information provision and research; early warning and disaster relief; facilitating adaptation options; regulating the distributional impacts of adaptation; and regulating infrastructures. The article concludes by arguing that the real integration problem for adaptation policy relates to how it is embedded in sectoral policies such as agriculture and transport, rather than how to achieve integration with mitigation policies.

Keywords: Climate change; Adaptation; Public policy

1. Introduction

All natural and social systems are, to a greater or lesser extent, adapted to the climates they experience. Climatic change imposes new pressures on those systems to adjust in response. In natural ecosystems these pressures will be experienced as new selection pressures, changing the structure and dynamics of populations. In social systems, these pressures will also be experienced as selection pressures, but in addition there will be scope for reflexivity, innovation and change, as people and organizations adjust to remain sustainable.

Many of these adjustments will be made privately, by individuals, households and businesses, and they are likely to yield principally private benefits. However, there are good reasons to believe that private adaptation, by itself, will remain at a level below what might be deemed socially or politically desirable (IPCC, 2001b). This is due to spillover effects (certain benefits of private adaptation

* Corresponding author. Tel.: +31-20-598-9525
E-mail address: frans.berkhout@ivm.vu.nl

may be shared inadvertently with others), uncertainty about the distribution of benefits and costs of adaptation, and the mismatch between the distribution of climate vulnerability and the capacity to adapt. These problems are manifested at local, regional, national, as well as international, levels. In addition, there will be a range of adjustments that need to take place in the public sphere. These include changes to major infrastructures, as well as changes in standards and regulations that will give private actors the framework and incentives to adapt. The need to respond to more rapid global environmental change may also influence patterns of national and international governance at a deeper level as well, as the value of diversity and flexibility in socio-technical systems grows.

For these reasons – that there will tend to be under- or maladaptation in the private sphere and because adjustments are necessary in the public sphere – there is a clear role for policy in motivating and shaping adaptation to climatic change. Although this was acknowledged in the IPCC Third Assessment Report (TAR), and provisions for adaptation exist within the Framework Convention on Climate Change (FCCC),[1] adaptation has until recently failed to be given the same attention as mitigation in the development of climate policy, within the EU and internationally. A mature climate policy needs to find a place for both mitigation and adaptation.

This article has three main aims:

1. To summarize some key ideas related to social and economic adaptation to the impacts of climatic change
2. To explain the similarities and differences that exist between mitigating climate change and adapting to its consequences
3. To set out a rationale for policy intervention related to climate adaptation.

The next section outlines evidence of climate change and impacts. This is followed by sections dealing with the questions of who and what adapts to climate change impacts. Alternative models of how social and economic adaptation may occur are then discussed, followed by a section comparing the similarities and differences between mitigation (the reduction of greenhouse gas emissions) and adaptation (adjustments made in response to climate change impacts, real or perceived). The final two sections attempt to set out a rationale for 'adaptation policy'. The primary scope of the article is Europe, with the aim of illustrating problems that are more generic.

2. Climate change impacts and adaptation

Scientific evidence is accumulating that the global climate is changing. Over the last century average surface temperatures have risen by 0.6°C, and the Intergovernmental Panel on Climate Change (IPCC) reported in 2001 that most of this warming over the past 50 years can be explained by increasing concentrations of greenhouse gases in the Earth's atmosphere (IPCC, 2001a). The IPCC projected that average global temperatures would rise by a further 1.4–5.8°C over the next century. Even if anthropogenic greenhouse gas emissions were ended now, temperatures would continue to rise for three to four decades, and sea levels for longer.

The main impacts of rising temperatures are expected to be rising average sea levels (by between about 10 and 90 cm), an accelerated hydrological cycle leading to increased precipitation, and the likelihood of greater climatic variability, including greater extremes of temperature, precipitation and storminess. Research suggests that these climatic changes will affect both natural systems and human activities. Indeed, there is already evidence from nature of responses to changing climate.

Examples include shrinkage of glaciers, thawing of permafrost, lengthening of mid- and high-latitude growing seasons, poleward shifts of plant and animal ranges, and earlier flowering of trees, emergence of insects and egg-laying birds (IPCC, 2001a, p. 3). We can see that climatic, terrestrial and marine systems are all being reshaped by a warmer climate, with multiple effects on human welfare.

Human and natural systems that are most strongly interwoven are also those that are most likely to be affected by climate change. These include: agriculture and food; forestry; freshwater resources; terrestrial and freshwater ecosystems; coastal zones and marine ecosystems; human settlement and industry; insurance and financial services; and human health. But impacts will be experienced beyond these sectors as well, to the extent that a changing climate may come to be seen as influencing most human activities, especially under more extreme warming scenarios. As with natural systems, we can already observe that societal actors are responding to the direct and indirect impacts of a changing climate. For instance, farmers in some temperate regions are taking advantage of longer growing seasons to increase crop yields, while water resource managers are building changing rainfall patterns into their forward plans. Such responses and adjustments are termed 'adaptation'.

Adaptation to social and environmental change is a feature of all human societies (Rayner and Malone, 1998). To a greater or lesser extent, all societies also have the capacity to adapt to a changing climate, including those experiencing a wide range of conditions from polar to desert climates. More recently, an argument has been made for considering the mutual interactions between human and environmental systems as the centrepiece of a 'sustainability science' (Turner et al., 2003). Sustainable societies exist across the whole range of the Earth's climates. Each of these environments has familiar seasonal patterns of weather, and these influence many aspects of economic and social life, including diet, dress and settlement. Societies are also accustomed to dealing with the inherent variability of climate, and have a capacity to cope with extremes and weather-related disasters.

It is important to recognize that adaptation to changes in climate will occur in the context of many other changes and adaptations, in both natural and socio-economic systems (IPCC, 2001b). For instance, while temperatures have been rising slowly over the past century, there have been dramatic changes in the exploitation of natural resources and in technological and social systems across the world. While changes in climate experienced over the past 50 years as a result of warming may have had impacts on human welfare and on the resilience of natural ecosystems, these are likely to have been experienced as background changes in the context of many other, more significant social, political and economic discontinuities. In future, the prospect is that climatic change will become more marked, with climate becoming a more significant driver of adaptive behaviour by people, organizations and countries, as well as challenging the capacity of these groups to adapt.

The crucial question is whether social and natural systems can change in response to a changing climate – implying both changes in mean conditions and in variability – and whether this can be achieved without suffering losses in overall social welfare or ecosystem functioning.[2] While more flexible and fast-changing aspects of social and natural systems will adapt relatively quickly and at low cost to a changing climate, more long-lived and inflexible features are likely to be more difficult and costly to adjust. In some parts of the world, whole economic sectors seem likely to be transformed. For instance, higher mean and peak temperatures around the Mediterranean may reduce its attractiveness as a summer holiday destination by the 2050s, while higher temperatures in northern Europe may make it more attractive (for review, see Maddison, 2001). The question is whether such a change, over the period projected, can be accomplished without major social costs and disruption. We should also expect surprises, with thresholds being crossed, and sudden, much more rapid, climatic shifts and social responses occurring as a result.

Given this background, there are several reasons why we need to understand more about adaptation (see IPCC, 2001b, p. 890, for a similar list):

1. Climate change cannot be totally avoided, and is likely to be continuous for many decades (in a transient state) and could be more rapid and pronounced than expected
2. The degree to which societal and natural systems are vulnerable to a changing climate will be influenced by whether they will or can adapt
3. Anticipatory adaptation is likely to be more effective and less costly than adaptation after the event
4. There are immediate social and economic benefits to be gained through better adaptation to climate variability and extreme events.

It is already clear that, on a global scale, the pattern of the impacts from climate change is likely to vary tremendously. For instance, some models suggest negative economic impacts in agriculture in tropical and sub-tropical regions of the world, even under low temperature-increase scenarios (Mendelsohn et al., 2000). In contrast, yields in mid-latitudes could increase with moderate temperature rises (IPCC, 2001a, p. 5–6). Even within the UK, while less precipitation is expected in the south-east by 2050, increased rainfall is projected for the north-west (Hulme et al., 2002). Likewise, the capacity to adapt will vary, with more affluent, knowledgeable and socially-cohesive societies perhaps being generally better able to respond (Adger, 2003). As the agricultural example shows, the greatest impacts may fall on societies that are currently least able to adapt effectively. We need to build our understanding of how these patterns of vulnerability and adaptive capacity intersect, and aim to focus our attention on these sectors and regions where the vulnerabilities are likely to be greatest.

3. Adapting to what?

In both natural and human systems, the range of environmental variability to which they are adapted has been termed the 'coping range' or resilience.[3] While it is often difficult to be precise about the coping range of an ecosystem, or of an organization, the general idea is useful in describing the discomfort, costs and risks that come with needing to cope with conditions that are outside common experience. The closer you get to the edge of the coping range, the greater will be the effort to maintain welfare or function. So, for example, in a wet summer, hotel owners need to work harder to attract customers, while still continuing to have a profitable business.

Four features of responses to climatic conditions stand out. First, climate is many-faceted – it is experienced as more than just a single distinct phenomenon, such as temperature or precipitation. Climatic conditions to which people or ecosystems respond and adapt are a combination of factors, and their particular effects on people are usually mediated by many other socio-economic factors. So, for instance, cold, wet and windy would be experienced differently than cold, wet and still, and would affect behaviours in particular ways, whether these are average conditions, or variable episodes. To give a more concrete example, precipitation experienced with higher wind speeds will influence the standards of water-tightness required in the construction of buildings – as is already being experienced in the southern UK – whereas higher precipitation alone might only have an influence on flood defence measures.

Second, the effects of varying climatic conditions (and on changes in these conditions) will differ across different social groups – a hot, dry summer may be good for ice-cream sales, but may be much less welcomed by some farmers.[4] This means that in talking about impacts, care needs to be taken about identifying what group or system is being affected, and what range of conditions are being considered (see Table 1). Not all coping ranges will be equivalent, even within a similar sector, species or ecosystem, so that any given change in climate is likely to produce both winners and losers.

Table 1. Climate-sensitive sectors and systems (adapted from Easterling et al., 2004, p. 3)

Socio-economic sectors	
Agriculture	High sensitivity and exposure, but high capacity to adapt
	EU agricultural output likely to rise up to 2–3°C temperature rise; and to fall beyond this
Forestry	High sensitivity, moderate capacity to adapt Substantial change in productivity and location of forests
Freshwater resources	High sensitivity, moderate to high capacity to adapt
	More droughts and floods, major infrastructural investment needed
Coastal zones	High sensitivity, variable capacity to adapt. Increased costs of sea defence
Built environment	Moderate sensitivity, variable capacity to adapt
	Locational change and climate-proofing of built environment costly
Tourism	Variable sensitivity, variable capacity to adapt. Possible changing seasonality of tourism across Europe
Natural systems	
Terrestrial ecosystems and freshwater aquatic ecosystems	High vulnerability Substantial change in distribution of species; loss of biodiversity expected
Coastal and marine ecosystems	High vulnerability Wetland areas substantially affected by changes in temperature and run-off

Third, social systems will adjust to the direct experience of climate change, but also to a host of indirect consequences that result from climatic changes. For example, the price of food commodities in the EU may vary as a result of harvest failures in another part of the world; or house-builders may no longer develop on fluvial floodplains because insurers are unwilling to insure properties in those areas. Fourth, climatic factors may have their greatest influence as sequences of events, as well as in the form of single catastrophic events. Natural and social systems often have the resilience to cope with single events, but become more vulnerable to the compounding effect of sequences of harmful events. A single event may place a system at the edge of its coping range; a follow-up event may push it outside this range. For example, water supply companies in the UK are able to cope with one or even two dry seasons. Water resources become seriously challenged in parts of the south and east if a dry summer and winter are followed by another dry summer (Arnell, 1998). We take as given a familiar level of variability of climate, and assume that natural and human systems are more or less adapted to them. Climate change will impose a different range of climatic conditions on natural and human systems, exposing them to new stresses as a result. It is also important to recognize that it is likely that these changes will be more or less continuous for a long period. Even if greenhouse gas emissions are stabilized at a level that prevents atmospheric concentrations from more than doubling pre-industrial levels, global climate will continue to become less similar to our current climate for the next century or so. Social systems therefore face not only a changed, but a continuously changing, climate.

4. Who and what adapts?

To illuminate the complex interactions between climate and the development of natural and social systems, analysts have developed some concepts useful for understanding adaptation (see Box 1). An important conclusion is that the vulnerability of a system is tied to its adaptive capacity. An adaptive system is likely to be less vulnerable than one which is less able to make adjustments that maintain productivity, functioning or welfare. Adaptation is therefore a way of reducing vulnerability to climate change. Effective adaptation will reduce the costs of damages experienced as a result of climatic change impacts, and will enable a system to take advantage of opportunities to improve performance that may arise from the changed conditions.[5]

Box 1: Adaptation concepts

Vulnerability: a measure of a system's susceptibility to climate change – a function of the system's exposure, sensitivity and adaptive capacity.

Exposure: the extent to which a climate-sensitive sector is in contact with climate.

Sensitivity: the degree to which a system is affected by climate change.

Adaptive capacity: how well a system can adjust to climatic changes to moderate potential damages (by changing exposure or sensitivity), to cope with the consequences of impacts (by recovering or maintaining welfare/system function in the face of climatic change) and to profit from new opportunities (assuming climate change affects social agents differentially).

Source: Adapted from IPCC (2001b).

Broadly speaking, there are three sides to adaptation (IPCC, 2001b):

1. Minimizing sensitivity or exposure to risk
2. Developing a capacity to cope after damages have been experienced
3. Acquiring the means to exploit new opportunities that arise.

In practice, adaptation may include a series of adjustments that attempt to strike a balance between these three broad objectives. For instance, the costs of reducing to very low levels exposures to climate-related risk may be prohibitive, therefore requiring some investment in contingency and recovery planning.

  Here the difference between natural and social systems needs to be more sharply drawn. For plants, animals and ecosystems, environmental changes impose new pressures that increase or decrease their ability to survive and reproduce. Their capacity to adapt will typically be quite limited. Biological systems are constantly responding to changing environmental conditions and to genetic variety. Over longer time periods they become redistributed and evolve. Gradual changes in conditions may be accommodated by natural ecosystems, but more rapid changes can be disruptive, especially in already stressed environments. Many of the world's ecosystems are already stressed by a variety of disturbances, including pollution, fragmentation and the invasion of exotic species (Easterling et al., 2004, p. 4). Climate change adds another stress.

People and organizations, in principle, have the capacity to make conscious and planned adjustments to the way they respond to the risk of climatic change. They can act to reduce their vulnerability and to make the most of advantageous changes in their environment. They can anticipate change, or they can respond to impacts, having assessed the alternatives. In general, adaptive capacity will be related to knowledge and awareness, access to resources, technology, social networks and attitudes to risk (Smit and Pilifosova, 2003). We would expect more knowledgeable, better-resourced, more equitable and more interconnected groups and organizations to have a greater range of adaptation options available to them, and to have a greater capacity to put these into practice. Likewise, we would expect those with a more precautionary attitude to adapt in anticipation of expected future impacts, while others prefer to 'muddle through' in reaction to experienced damages or opportunities.

Early research on adaptation to climate change impacts has tended to make the broad assumption that the greatest vulnerabilities would be in the developing world, rather than in Western industrialized countries. This was partly because global climate models predict that some of the most marked changes in climate will be in tropical (and polar) regions, but also because less-wealthy societies are seen as more vulnerable in the face of many other economic, social and environmental changes. To give a simple example, while low-lying regions of Europe may be defensible against sea level rise, it is likely that some Pacific island states will not be habitable by the middle of this century. This perception is now changing. Not only is it recognized that relatively less-developed societies may have considerable adaptive capacities based on indigenous knowledge, societal ties and networks (Mortimore, 1989), but it has also become clearer that more-developed societies may have considerable vulnerability to climate change, partly as a result of being more closely linked with the rest of the world through global trade and investment. Tightly coupled technological and economic systems in the industrialized world may have a 'brittleness' to some impacts of climate change. Adaptation is therefore a serious issue in Europe as well, and needs to be part of a response to climatic change as much as are efforts at mitigating climatic change through reducing emissions of greenhouse gases to the atmosphere.

5. Adaptation and mitigation

In developing an integrated climate policy, which includes provisions for adaptation to climate change impacts, as well as mitigation, it is important to relate the two core objectives to one another. At an aggregate level, mitigation and adaptation can be viewed as being partial substitutes for each other. For instance, less effort on mitigation – by aiming for a higher stabilization level for atmospheric CO_2 concentrations – will imply a greater effort on adaptation over the longer-term future. Part of the justification by some countries for not incurring high mitigation costs in the short term draws on the expectation that the costs of adaptation in the medium and long term will be lower. This is because overall welfare will have improved with economic growth, and because technological (and institutional) change is assumed overtime to make adaptation relatively easier and cheaper. Conversely, one of the arguments for accepting the costs of substantial mitigation efforts in the short term is based on the recognized uncertainty that exists around the costs of adaptation to more rapid and damaging climatic change. While the EU has committed itself to a target of a less than 2°C global temperature increase above pre-industrial levels in its climate policy, this also implies a certain quotient of climate change (CEC, 2005). This target implicitly represents a trade-off between mitigation and adaptation over the longer term. There is also an important international political assumption underlying this construction. In very broad terms, under the FCCC, mitigation is to be carried out by more-developed countries, while less-developed countries are promised assistance both with pursuing less carbon-intensive development paths, and with adaptation to climate change impacts.[6]

Strategically, it is therefore clear that mitigation and adaptation are bound together, at least over the long term. This has led some commentators to consider the potential for synergies between the practical implementation of adaptation and mitigation (Wilbanks et al., 2003). While there are likely to be many opportunities for linkages between mitigation and adaptation actions, it is also important to recognize that some basic features of the two objectives are divergent. Indeed, the 'problem structure' of adaptation appears significantly different from that of mitigation.

First, while most mitigation will bring 'common good' benefits (typically at an international level), the benefits of adaptation actions will often be private or localized. Mitigation investments in renewable power generation capacity will contribute to lower atmospheric concentrations of carbon dioxide over the long term, with gains in terms of reduced global climate change. Adaptation investments, even in major new infrastructures such as raised sea defences, will bring benefits only to settlements and ecosystems protected directly by them. Many adaptation actions will be at a more local scale, and will be implemented because they are expected to generate mainly private benefits.

Second, the benefits of mitigation will typically be experienced over the long run, since amelioration of climatic change through reductions in greenhouse gas emissions will typically occur over decadal time-scales. On the other hand, benefits from adaptation may become apparent over the short run, since they may be in response to already experienced changes in climate. The time-profiles of mitigation and adaptation are therefore often likely to be difficult to reconcile.

Third, while adaptation is concerned with multiple adjustments related to manifold direct and indirect interactions between climate and human activities (and natural ecosystems), mitigation is concerned with the relatively more bounded problem of anthropogenic greenhouse gas emissions. Early research on adaptation shows that the assessment of climate vulnerability, as with many risk assessments, can often be an open-ended process involving many aspects of individual and organizational activities (Berkhout et al., 2004). This complexity of ten allows for a wide range of adjustment options related to many aspects of these activities. Table 2 gives a summary picture of adaptation options available, in principle, to UK house-builders. Many of these represent extensions of conventional practices and innovations in the sector (Hertin et al., 2003). While mitigation can be achieved through technological as well as behavioural means, the measure of effectiveness is unitary – lower greenhouse gas emissions. Reduced vulnerability (or greater resilience) will tend to be multidimensional, including and usually hard to measure at local or broader scales (Adger et al., 2004).

Finally, while the energy sector will be the focus of mitigation, adaptation to the impacts of climate change will be occurring in a number of different sectors (and ecosystems), many of them not substantial contributors to greenhouse gas emissions. The major adaptations to climatic change expected in the energy sector are those linked to climate mitigation policy. Having said this, many opportunities for synergy between mitigation and adaptation exist and these need to be exploited, where practicable.

Table 2. Adaptation measures in the UK house-building sector

Function	Commercial adaptation / business model	Technological adaptation	Financial adaptation	Information and monitoring
Buying Land	• avoid areas at risk from flooding and erosion • learn to manage flood risk	• improved use of decision support tools (GIS etc)	• costing in potential climatic effects increased option buying	• monitor climate change impacts on land prices
Designing Houses	• provide 'climate-proofing' options	• use higher standards and new materials • stronger foundations • designs suitable for offsite manufacture	• link mortgage conditions to climate-proof building design	• monitor climate change impacts on buildings
Building Houses	• increase flexibility of construction process • improve supply chain management	• off-site manufacture use of weather-resistant techniques	• insure building-sites against weather damage	• monitor weather-induced conditions on building sites
Selling Houses	• sales strategies taking account of climate changes issues	• offer additional anti-flooding or storm proofing options	• improve buildings insurance against weather damage	• monitoring customer perceptions in relation to climate change issue
Maintaining Houses	• move away from maintenance (e.g. sub-contract)	• retro-fit new technologies (e.g. improved roofs)	• financial reserve for maintenance costs (housing associations) • restrict and shorten warranty	• monitoring of climate change impacts on maintenance

6. Models of adaptation

Many accounts of adaptation in response to climate change have made assumptions about when and whether people will adapt. For instance, some assessments hold that people and organizations may not adapt at all, but will continue to operate as if nothing had changed (the so-called 'naive' or 'dumb farmer' assumption; IPCC, 2001a, p. 887; Tol et al., 1998). Other assessments question whether people will adapt to the anticipated impacts of climate change, and assert that they will react only once evidence of damage (or opportunity) exists (Mendelsohn et al., 2000). Still others assume various levels of adaptation (Rosenzweig and Parry, 1994). While reactive adaptation has the benefit of happening in the context of less uncertainty (you know more about what you are responding to and what the benefits of adaptive action may be), there may be additional benefits from early action, such as the avoidance of certain losses.

Such assumptions are key to models that have tried to calculate the economic costs and benefits of adaptation (as compared with 'no adaptation' scenarios). The 'dumb farmer' case gives a 'worst case' scenario for the costs and damages associated with climatic change, while an anticipatory strategy is often held to give a 'best case' scenario. In all these assessments there often continues to be considerable uncertainty about the precise nature of possible impacts, about how vulnerability will be expressed and experienced, and also about the adaptation measures that may be taken by people and organizations, particularly in the medium- and long-term future. Assumptions about how people will act, and about the economic consequences of these actions, are one way of coping with the limited evidence that exists about future social responses to climatic changes. While there are historical analogues to draw on, great care is needed in the lesson-learning for the present day (Meyer et al., 1998). What economic analyses do express is the idea that adaptation is likely to occur only when there is a perceived advantage to those who are adapting.

An alternative, more bottom-up, approach begins from the position of the people, organizations and institutions that are (or will be) adapting (i.e. adapting agents). One of the findings of this research is that the actual and perceived vulnerability of adapting agents can vary a great deal, even in apparently similar contexts. For instance, among UK water companies there are important regional differences in the vulnerability to changing precipitation patterns as a result of climate change (Berkhout et al., 2004). In the north-west of England, water resources are likely to be put under far less stress than in the south-east where, with a rising population and limited surface storage capacity, there are already problems in matching supply with demand in some areas. Likewise, the capacity to adapt in order to reduce exposure to climate impacts, and to build resilience to cope with impacts, can vary considerably. A hotel by the beach in Brighton on the south coast of England faces a different profile of climate risks than a hotel located 1 km inland, and the measures that might be necessary to build resilience (by taking out extended insurance cover against storm damage, for instance) are likely to be more costly for one than the other.

This means that, unlike policies and actions on climate mitigation, policies and actions related to adaptation often need to be sensitive to features at the micro-level of individuals, households, businesses and localities. Vulnerability to climate change may be universal, as is the capacity to adapt, but the gradients of vulnerability and adaptive capacity that exist between adapting agents tend to be steep. This variability, as well as the uncertainty that exists about the potential value of adaptation, means that it is very difficult to argue for a 'best' adaptation strategy for any given adapting agent. Even where there is general awareness of climate change and its possible consequences, some agents will choose to adapt and will employ a range of strategies, while others will not. Both may be appropriate and well-founded responses in the context of uncertainty. As more is known that helps us to evaluate climate vulnerability and the benefits of adaptation, we would expect greater convergence in adaptive behaviour.

7. Key problems in adaptation

If adaptation to variable environmental conditions is normal for all natural and social systems, and we can observe adaptive responses to changing conditions, it is not immediately obvious that there should be a role for government. We might take the view that natural systems – short of creating artificial climates around them – are for the most part difficult to protect against climatic changes. Likewise, if private actors can be relied upon to act in their own best interests, and the best adaptation is after-the-fact, then perhaps social learning and the market should be seen as the basis for adaptive responses. Indeed, by protecting people and organizations from the effects of change, some economists would argue that governments run the risk of encouraging maladaptation.[7]

On the other hand, such an analysis seems insufficient. Government is likely to play a number of roles in enabling, influencing and implementing adaptation to climatic change. At the highest level, the EU and Member State governments have a role in determining the balance between mitigation and adaptation, as part of an integrated climate policy. But there is a range of other roles specific to adaptation that governments can play, working independently and together through the EU and the international system. To understand what these roles will be, we need to highlight some key problems that have been identified in adaptation research so far.

7.1. Awareness of climate vulnerability

Understanding the exposure, sensitivity and adaptedness of a natural or human system to future changing climate is complex. Vulnerability assessment is a growing field and needs to provide practical tools and applications that can be used by people and organizations, and in the management of natural ecosystems. At the root of vulnerability assessment and management must be improved predictions of climatic change and impacts, especially over the short- to medium-term future. Private actors generally will not pay for the science needed for the assessment and management of vulnerability.

7.2. Awareness of adaptation options

Although organizations are continually under pressure to change the way they do things, climate risk has been integrated into innovation processes in only a few cases. This is at least partly because organizations are not yet aware of the measures (technological, institutional and so on) that could be taken by them to moderate these climate vulnerabilities and risks. Adaptation needs to be integrated more widely, and may involve, over the short run, only small adjustments to the procedures of many organizations.

7.3. Uncertainty and motivation

For many organizations there will continue to be considerable uncertainty about the precise nature and risks of changing climate and variability, about their climate vulnerability and about the benefits of adaptation. To some extent this uncertainty will remain irreducible, but there can be a role for better climate prediction and more tailored information, especially for smaller organizations. Anticipatory adaptation will become more likely with lower uncertainties about impacts and benefits of adaptation. There may also be collective, broad-scale benefits from adaptation which cannot be captured if private actors are not informed about and given incentives to adapt.

7.4. Adaptation spillovers

As with many forms of innovation and change, the benefits of adaptation to climatic changes may not be appropriated entirely by the agent making the change. There may be other beneficiaries from the actions, knowledge and experience that an innovator has invested in. These 'spillovers' can lead systematically to a collective under-investment in adaptation, generating a rationale for policy and legal interventions, much as with intellectual property rights and patenting regimes. In addition, climate adaptations that reduce the vulnerability of one agent may generate either negative or positive consequences for others. As we have seen, vulnerability to climate change is likely to be unequally distributed across different groups in society, nationally and internationally. In general, we would expect some groups and societies to be able to moderate their vulnerability more effectively, often through transferring risks onto others. Adaptation may therefore lead to a deepening of already existing inequities. There is a role for policy to protect both the innovative and the vulnerable.

7.5. Constraints on adaptation

Much adaptation will draw on resources (including capital, knowledge, technology and consent) that are not held by the adapting agents themselves. While some of the resources will be made available through the market, there are also likely to be scarcities and constraints – partly as a result of the problems of awareness discussed above. Policy has a role in modifying and perhaps removing some of these regulatory, market or infrastructural constraints that exist to adaptation. It is also likely that in giving adapting agents greater scope to adapt (extending their so-called 'adaptation space') new conflicts will be generated with other environmental, social or economic objectives. There may be 'win–wins', but we should also expect trade-offs, especially where new resources are required to modify vulnerability or improve adaptive capacity.

8. Roles for policy in adaptation

This assessment of some key challenges facing social and economic agents adapting to climate change, provides a structure for discussing potential roles for government policy. As we have argued, adaptation needs to become a substantial and integral component of climate policy at the EU, Member State and sectoral levels. A variety of rationales have been given for public policy related to climate adaptation. Klein and Tol (1997) argue that public policy related to adaptation should have four objectives: increasing robustness of infrastructures; increasing flexibility and adaptability of vulnerable managed systems; reversing trends that increase vulnerability; and improving awareness and preparedness.

Drawing on these insights and on the arguments above, we suggest that the primary objectives for government action could be: to inform the potentially vulnerable; to assist in the provision of disaster relief; to incentivize and enable adaptation; to regulate adaptation 'spillovers' and risk-shifting; and to plan and regulate long-term and infrastructural assets so as to reduce future vulnerabilities.

8.1. Information, knowledge and learning

Governments have played a major role in the sponsorship of climate science and in the provision of tools such as global, regional and national climate scenarios. This informational role is being continually expanded. Experience shows that awareness of climate impacts and vulnerability assessment remains patchy, being well-developed in some sectors, such as water services and insurance, and generally poor in many other sectors.

8.2. Early-warning and disaster relief

Most governments have in place plans, organizations and resources to alert people to weather-related disasters and to cope with the consequences, at home and abroad. These will need to be continually reviewed as the frequency, scope and intensity of weather-related disasters changes as a result of climate change. An important aspect of adaptive capacity is the capacity to cope with weather-related events.

8.3. Facilitating adaptation options, guiding adaptation and enabling adaptive capacity

There are strong 'public good' arguments for investing in scientific and technological resources that may be widely adopted in response to climate change. A standard response to greater uncertainty is to broaden the portfolio of adaptations that are available to vulnerable sectors. Beyond investing in innovations that may be applied by adaptors, there is also a clear role for regulators to signal the need to adapt to the private sector. The rationale for this is the potential for under-investment in adaptation by economic actors confronted by high uncertainty about the likelihood and consequences of climate change impacts.

8.4. Regulating distributional consequences of adaptation

Unregulated, it is likely that the most vulnerable social groups will end up bearing many of the new social and economic risks that arise as a result of climate change. A simple example of this is the proposed reduction in the term (from 3 years to 2 years) of liability insurance covering new houses in the UK, partly as a response to heightened risks of storm damage (Hertin et al., 2003). In this way the house-owner, rather than the house-builder's insurer, comes to take on an increased risk.

8.5. Infrastructure planning and development

Water, transport and energy infrastructures are likely to be influenced by changing climate, as is the distribution of settlements, especially in coastal and fluvial flood plains. Modification of infrastructures and of spatial plans in response to experienced and predicted climate impacts is another area in which Governments will play a major role. Difficult trade-offs are likely to be necessary between conflicting social, economic and environmental objectives as a result.

9. Conclusion

We have argued that adaptation needs to become a central feature of climate policy, having an independent status equivalent to greenhouse gas emissions reduction at the national, regional and local levels. If climatic changes are already observable, then adaptation is also likely to be occurring, and given the inertia in the global climate system, will continue to unfold over periods of several decades, no matter what is achieved on emissions reductions. Analysis of the costs and feasibility of adaptation will also come to underpin the politics of mitigation policy, by showing why mitigation is economically and politically necessary.

This article has sought to outline some of the main problems to which 'adaptation policy' could be directed. These include the problems of awareness and uncertainty that will militate against adaptation by people, businesses and other organizations, and the problems of sharply differentiated knowledge, vulnerability and economic spillovers that need to be managed collectively. In many cases, these are new expressions of well-understood economic and risk governance problems. They require attention because climatic change is a novel environmental problem to which policy systems are themselves not yet well attuned.

We have also sought to make an argument in relation to the question of climate policy integration. While at a strategic, though perhaps largely theoretical, level there are choices to be made between mitigation and adaptation, others have argued that, in the implementation of climate policy, mitigation and adaptation need to be integrated. We have argued that while there may be potential for synergy there are differences in the problem-structure of mitigation and of adaptation. It may therefore be more helpful to avoid eliding the two, and to start instead with the presumption that they are separate domains. This also counts for the conduct of policy in relation to climate adaptation. While there may be some domains, such as disaster relief or infrastructure development, where it may be useful to think of a distinct field of policy action termed 'climate adaptation', in many other fields it would be more efficient and effective to seek to build adaptation measures into existing processes of policy analysis, implementation and evaluation in sectors that may not directly address climate as an issue. This means that adaptation policy will for a large part be adjustments in other policy domains, including agriculture, transport, water resource management, trade, science, technology and innovation (STI) and so on. This, of course, poses special challenges for policy development and coordination. As climate adaptation becomes a mainstream feature of climate policy, so the question of how far it can be integrated and how far it needs to stand alone, will need continue to be asked.

Acknowledgements

I would like to thank participants at the European Forum for Integrated Environmental Assessment (EFIEA) Workshop held in Den Haag on 30–31 August 2004 and three referees for their useful comments.

Notes

1   Article 2 argues that greenhouse gas concentrations need to be stabilized and that '…a level should be achieved within a time-frame sufficient to allow ecosystems to adapt naturally to climate change, to ensure that food production is not threatened and to enable economic development to proceed in a sustainable manner'.

2   Note that this does not assume that adaptation is always at the service of maintaining the *status quo*. 'All socio-economic systems are continually in a state of flux in responding to changing circumstances, including climatic conditions' (IPCC, 2001a, p. 889).

3   Formally, coping range represents the range of critical environmental variables (including for instance, temperature or the price of a factor of production) across which a system can operate without loss of performance (Hewitt and Burton, 1971; Fukui, 1979). Once the characteristic degree of variability of these critical variables is exceeded, the system has to expend new resources to sustain characteristic levels of integrity and performance. A related concept is 'resilience'. Folke et al. (2002, p. 13) define this as '…the capacity to absorb shocks while maintaining function'. A difference may be that the resilience concept does not contain a definable limit to the capacity to absorb shocks, as implied in the notion of a 'range'. Resilience is a more graded concept in which 'thresholds' (Parry, 1986) or 'bands of tolerance' (Warrick et al., 1986) do not necessarily feature.

4   The hot dry summer of 2003 was responsible for major agricultural yield reductions in large parts of southern and eastern Europe.

5   US studies report that adaptation measures in agriculture may lead to reductions in adverse costs of climate change of between 29% and 60% (Tol et al., 1998).

6   The National Adaptation Programmes of Action (NAPA) are the clearest articulation of this support by the global North for adaptation in the global South.

7   An example of this might be the public funding of contingency planning in vulnerable regions, such as coasts exposed to intense tropical storms, which tends to encourage settlement in those regions.

References

Adger, W.N., 2003. Social capital, collective action and adaptation to climate change. Economic Geography 79(4), 387–404.

ADGER, N.N., Brooks, N., Kelly, M., Bentham, S. and Eriksen, S., 2004. New indicators of vulnerability and adaptive capacity. Tyndall Centre Technical Report 7, UEA, Norwich.

Arnell, N.W., 1998. Climate Change and Water Resources in Britain. Climatic Change 39: 83–110.